FE CIVIL PRACTICE PROBLEMS

for the Civil Fundamentals of Engineering Exam

Michael R. Lindeburg, PE

The Power to Pass®
www.ppi2pass.com

Professional Publications, Inc. • Belmont, California

Report Errors and View Corrections for This Book

Everyone benefits when you report typos and other errata, comment on existing content, and suggest new content and other improvements. You will receive a response to your submission; other engineers will be able to update their books; and, PPI will be able to correct and reprint the content for future readers.

PPI provides two easy ways for you to make a contribution. If you are reviewing content on **feprep.com**, click on the "Report a Content Error" button in the Help and Support area of the footer. If you are using a printed copy of this book, go to **ppi2pass.com/errata**. To view confirmed errata, go to **ppi2pass.com/errata**.

FE Civil Practice Problems

Current printing of this edition: 1

Printing History

edition number	printing number	update
1	1	New book.

Copyright © 2014 Professional Publications, Inc. All rights reserved.

All content is copyrighted by Professional Publications, Inc. (PPI). All rights reserved. No part, either text or image, may be used for any purpose other than personal use. Reproduction, modification, storage in a retrieval system or retransmission, in any form or by any means, electronic, mechanical, or otherwise, for reasons other than personal use, without prior written permission from the publisher is strictly prohibited. For written permission, contact PPI at permissions@ppi2pass.com.

Printed in the United States of America.

PPI
1250 Fifth Avenue
Belmont, CA 94002
(650) 593-9119
ppi2pass.com

ISBN: 978-1-59126-440-8

Library of Congress Control Number: 2014931692

Topics

Mathematics

Probability/ Statistics

Fluid Mechanics

Hydraulics/ Hydrologic Sys.

Environmental Engineering

Geotechnical Engineering

Statics

Dynamics

Where do I find help solving these Practice Problems?

FE Civil Practice Problems presents complete, step-by-step solutions for more than 460 problems to help you prepare for the Civil FE exam. You can find all the background information, including charts and tables of data, that you need to solve these problems in the *FE Civil Review Manual*.

The *FE Civil Review Manual* may be obtained from PPI at **ppi2pass.com** or **feprep.com**, or from your favorite print book retailer.

Table of Contents

Preface

The purpose of this book is to prepare you for the National Council of Examiners for Engineering and Surveying (NCEES) fundamentals of engineering (FE) exam.

In 2014, the NCEES adopted revised specifications for the exam. The council also transitioned from a paper-based version of the exam to a computer-based testing (CBT) version. The FE exam now requires you to sit in front of a monitor, respond to questions served up by the CBT system, access an electronic reference document, and perform your scratch calculations on a reusable, no-erase notepad. You may also use an on-screen calculator with which you will likely be unfamiliar. The experience of taking the FE exam will probably be unlike anything you have ever, or will ever again, experience in your career. Similarly, preparing for the exam will be unlike preparing for any other exam.

The CBT FE exam presented three new challenges to me when I began preparing instructional material for it. (1) The subjects in the testable body of knowledge are oddly limited and do not represent a complete cross section of the traditional engineering fundamentals subjects. (2) The NCEES *FE Reference Handbook* (*NCEES Handbook*) is poorly organized, awkwardly formatted, inconsistent in presentation, and idiomatic in convention. (3) Traditional studying, doing homework while working towards a degree, and working at your own desk as a career engineer are poor preparations for the CBT exam experience.

No existing exam review book overcomes all of these challenges. But, I wanted you to have something that does. So, in order to prepare you for the CBT FE exam, this book was designed and written from the ground up. In many ways, this book is as unconventional as the exam.

This book covers all of the knowledge areas listed in the NCEES Civil FE exam specifications. And, with the exceptions listed in "How to Use This Book," for better or worse, this book duplicates the terms and variables of the *NCEES Handbook* equations.

NCEES has selected, what it believes to be, all of the engineering fundamentals important to an early-career, minimally-qualified engineer, and has distilled them into its single reference, the *NCEES Handbook*. Personally, I cannot accept the premise that engineers learn and use so little engineering while getting their degrees and during their first few career years. However, regardless of whether you accept the NCEES subset of engineering fundamentals, one thing is certain: In serving as your sole source of formulas, theory, methods, and data

during the exam, the *NCEES Handbook* severely limits the types of questions that can be included in the FE exam.

The obsolete paper-based exam required very little knowledge outside of what was presented in the previous editions of the *NCEES Handbook*. That *NCEES Handbook* supported a plug-and-chug examinee performance within a constrained body of knowledge. Based on the current FE exam specifications and the *NCEES Handbook*, the CBT FE exam is even more limited than the old paper-based exam. The number (breadth) of knowledge areas, the coverage (depth) of knowledge areas, the number of questions, and the duration of the exam are all significantly reduced. If you are only concerned about passing and/or "getting it over with" before graduation, these reductions are all in your favor. Your only deterrents will be the cost of the exam and the inconvenience of finding a time and place to take it.

Accepting that "it is what it is," I designed this book to guide you through the exam's body of knowledge.

I have several admissions to make: (1) This book contains nothing magical or illicit. (2) This book, by itself, is only one part of a complete preparation. (3) This book stops well short of being perfect. What do I mean by those admissions?

First, this book does not contain anything magical. It's called a "practice problems" book, and though it will save you time in assembling hundreds of practice problems for your review, it won't learn the material for you. Merely owning it is not enough. You will have to put in the "practice" time to use it.

Similarly, there is nothing clandestine or unethical about this book. It does not contain any actual exam questions. It was written in a vacuum, based entirely on the NCEES Civil FE exam specifications. This book is not based on feedback from actual examinees.

Truthfully, I expect that many exam questions will be similar to the questions I have used because NCEES and I developed content with the same set of constraints. (If anything, NCEES is even more constrained when it comes to fringe, outlier, eccentric, original topics.) There is a finite number of ways that questions about Ohm's law ($V = IR$) and Newton's second law of motion ($F = ma$) can be structured. Any similarity between questions in this book and questions in the exam is easily attributed to the limited number of engineering formulas and concepts, the shallowness of coverage, and the need to keep the entire solution process (reading, researching, calculating, and responding) to less than three minutes for each question.

Let me give an example to put some flesh on the bones. As any competent engineer can attest, in order to calculate the pressure drop in a pipe network, you would normally have to (1) determine fluid density and viscosity based on the temperature, (2) convert the mass flow rate to a volumetric flow rate, (3) determine the pipe diameter from the pipe size designation (e.g., pipe schedule), (4) calculate the internal pipe area, (5) calculate the flow velocity, (6) determine the specific roughness from the conduit material, (7) calculate the relative roughness, (8) calculate the Reynolds number, (9) calculate or determine the friction factor graphically, (10) determine the equivalent length of fittings and other minor losses, (11) calculate the head loss, and finally, (12) convert the head loss to pressure drop. Length, flow quantity, and fluid property conversions typically add even more complexity. (SSU viscosity? Diameter in inches? Flow rate in SCFM?) As reasonable and conventional as that solution process is, a question of such complexity is beyond the upper time limit for an FE question.

To make it possible to be solved in the time allowed, any exam question you see is likely to be more limited. In fact, most or all of the information you need to answer a question will be given to you in its question statement. If only the real world were so kind!

Second, by itself, this book is inadequate. It was never intended to define the entirety of your preparation activity. While it introduces problems covering essentially all of the exam knowledge areas and content in the *NCEES Handbook*, an introduction is only an introduction. To be a thorough review, this book needs augmentation.

By design, this book has four significant inadequacies.

1. This book is "only" 238 pages long, so it cannot contain enough of everything for everyone. The number of practice problems that can fit in it are limited. The number of questions needed by you, personally, to come up to speed in a particular subject may be inadequate. For example, how many questions will you have to review in order to feel comfortable about divergence, curl, differential equations, and linear algebra? (Answer: Probably more than are in all of the books you will ever own!) So, additional exposure is inevitable if you want to be adequately prepared in every subject.

2. This book does not contain the *NCEES Handbook*. This book is limited in helping you become familiar with the idiosyncratic sequencing, formatting, variables, omissions, and presentation of topics in the *NCEES Handbook*. The only way to remedy this is to obtain your own copy of the *NCEES Handbook* (available in printed format from PPI and as a free download from the NCEES website) and use it in conjunction with your review.

3. This book is not a practice examination (mock exam, sample exam, etc.). With the advent of the CBT format, any sample exam in printed format is little more than another collection of practice questions. The actual FE exam is taken sitting in front of a computer using an online reference book, so the only way to practice is to sit in front of a computer while you answer questions. Using an online reference is very different from the work environment experienced by most engineers, and it will take some getting used to.

4. This book does not contain explanatory background information, including figures and tables of data. Though all problems have associated step-by-step solutions, these solutions will not teach you the underlying engineering principles you need to solve the problems. Trying to extrapolate engineering principles from the solutions is like reading the ending of a book and then trying to guess at the "whos, whats, wheres, whens, and hows." In other words, reviewing solutions is only going to get you so far if you don't understand a topic. To truly understand how to solve practice problems in topics you're unfamiliar with, you'll need an actual review manual like the one PPI publishes, the *FE Civil Review Manual*. In it, you'll find all the "whos and whats" you were previously missing and these problems' "endings" will make much more sense.

Third, and finally, I reluctantly admit that I have never figured out how to write or publish a completely flawless first (or, even subsequent) edition. The PPI staff comes pretty close to perfection in the areas of design, editing, typography, and illustrating. Subject matter experts help immensely with calculation checking. And, beta testing before you see a book helps smooth out wrinkles. However, I still manage to muck up the content. So, I hope you will "let me have it" when you find my mistakes. PPI has established an easy way for you to report an error, as well as to review changes that resulted from errors that others have submitted. Just go to **ppi2pass.com/errata**. When you submit something, I'll receive it via email. When I answer it, you'll receive a response. We'll both benefit.

Best wishes in your examination experience. Stay in touch!

Michael R. Lindeburg, PE

Acknowledgments

Developing a book specific to the computerized Civil FE exam has been a monumental project. It involved the usual (from an author's and publisher's standpoint) activities of updating and repurposing existing content and writing new content. However, the project was made extraordinarily more difficult by two factors: (1) a new book design, and (2) the publication schedule.

Creating a definitive resource to help you prepare for the computerized FE exam was a huge team effort, and PPI's entire Product Development and Implementation (PD&I) staff was heavily involved. Along the way, they had to learn new skills and competencies, solve unseen technical mysteries, and exercise professional judgment in decisions that involved publishing, resources, engineering, and user utility. They worked long hours, week after week, and month after month, often into the late evening, to publish this book for examinees taking the exam.

PPI staff members have had a lot of things to say about this book during its development. In reference to you and other examinees being unaware of what PPI staff did, one of the often-heard statements was, "They will never know."

However, I want you to know, so I'm going to tell you.

Editorial project managers Chelsea Logan, Magnolia Molcan, and Julia White managed the gargantuan operation, with considerable support from Sarah Hubbard, director of PD&I. Christina Gimlin, senior project manager, cut her teeth on this project. Production services manager Cathy Schrott kept the process moving smoothly and swiftly, despite technical difficulties that seemed determined to stall the process at every opportunity. Christine Eng, product development manager, arranged for all of the outside subject matter experts that were involved with this book. All of the content was eventually reviewed for consistency, PPI style, and accuracy by Jennifer Lindeburg King, associate editor-in-chief.

Though everyone in PD&I has a specialty, this project pulled everyone from his or her comfort zone. The entire staff worked on "building" the chapters of this book from scratch, piecing together existing content with new content. Everyone learned (with amazing speed) how to grapple with the complexities of XML and MathML while wrestling misbehaving computer code into submission. Tom Bergstrom, technical illustrator, and Kate Hayes, production associate, updated existing illustrations and created new ones. They also paginated and made corrections. Copy editors Tyler Hayes, Scott Marley, Connor Sempek, and Ian A. Walker copy edited, proofread, corrected, and paginated. Copy editors Alexander Ahn, Manuel Carreiro, and Hilary Flood proofread and corrected. Scott's comments were particularly insightful. Staff interns Nicole Evans, EIT; Prajesh Gongal, EIT; and Jumphol Somsaad assisted with content selection, problem writing, and calculation checking. Staff interns Jeanette Baker, EIT; Scott Miller, EIT; Alex Valeyev, EIT; and Akira Zamudio, EIT, remapped existing PPI problems to the new NCEES Civil FE exam specifications.

Paying customers (such as you) shouldn't have to be test pilots. So, close to the end of process, when content was starting to coalesce out of the shapelessness of the PPI content management system, several subject matter experts became crash car dummies "for the good of engineering." They pretended to be examinees and worked through all of the content, looking for calculation errors, references that went nowhere, and logic that was incomprehensible. These engineers and their knowledge area contributions are: C. Dale Buckner, PhD, PE, SECB (Statics and Structural Design); John C. Crepeau, PhD, PE (Dynamics; Mathematics; and Mechanics of Materials); Joshua T. Frohman, PE (Computational Tools; Construction; Probability and Statistics; and Transportation Engineering); David Hurwitz, PhD (Computational Tools; Construction; Probability and Statistics; and Transportation Engineering); David Johnstone, PhD, PE (Environmental Engineering; Geotechnical Engineering; and Hydraulics and Hydrologic Systems); Liliana M. Kandic, PE (Fluid Mechanics; Statics; and Structural Design); Aparna Phadnis, PE (Engineering Economics; Environmental Engineering; Geotechnical Engineering; and Hydraulics and Hydrologic Systems); David To, PE (Dynamics; Fluid Mechanics; Mathematics; and Mechanics of Materials); and L. Adam Williamson, PE (Fluid Mechanics; Dynamics; and Materials).

Consistent with the past 36 years, I continue to thank my wife, Elizabeth, for accepting and participating in a writer's life that is full to overflowing. Even though our children have been out on their own for a long time, we seem to have even less time than we had before. As a corollary to Aristotle's "Nature abhors a vacuum," I propose: "Work expands to fill the void."

To my granddaughter, Sydney, who had to share her Grumpus with his writing, I say, "I only worked when you were taking your naps. And besides, you hog the bed!"

Thank you, everyone! I'm really proud of what you've accomplished. Your efforts will be pleasing to examinees and effective in preparing them for the Civil FE exam.

Michael R. Lindeburg, PE

How to Use This Book

This book is written for one purpose, and one purpose only: to get you ready for the FE exam. Because it is a practice problems book, there are a few, but not many, ways to use it. Here's how this book was designed to be used.

GET THE NCEES *FE REFERENCE HANDBOOK*

Get a copy of the NCEES *FE Reference Handbook* (*NCEES Handbook*). Use it as you solve the problems in this book. The *NCEES Handbook* is the only reference you can use during the exam, so you will want to know the sequence of the sections, what data are included, and the approximate locations of important figures and tables in the *NCEES Handbook*. You should also know the terminology (words and phrases) used in the *NCEES Handbook* to describe equations or subjects, because those are the terms you will have to look up during the exam.

The *NCEES Handbook* is available both in printed and PDF format. The index of the print version may help you locate an equation or other information you are looking for, but few terms are indexed thoroughly. The PDF version includes search functionality that is similar to what you'll have available when taking the computer-based exam. In order to find something using the PDF search function, your search term will have to match the content exactly (including punctuation).

There are a few important differences between the ways the *NCEES Handbook* and this book present content. These differences are intentional for the purpose of maintaining clarity and following PPI's publication policies.

- *pressure:* The *NCEES Handbook* primarily uses P for pressure, an atypical engineering convention. This book always uses p so as to differentiate it from P, which is reserved for power, momentum, and axial loading in related chapters.

- *velocity:* The *NCEES Handbook* uses v and occasionally Greek nu, ν, for velocity. This book always uses v to differentiate it from Greek upsilon, v, which represents specific volume in some topics (e.g., thermodynamics), and Greek nu, ν, which represents absolute viscosity and Poisson's ratio.

- *specific volume:* The *NCEES Handbook* uses v for specific volume. This book always uses Greek upsilon, v, a convention that most engineers will be familiar with.

- *units:* The *NCEES Handbook* and the FE exam generally do not emphasize the difference between pounds-mass and pounds-force. "pounds" ("lb") can mean either force or mass. This book always distinguishes between pounds-force (lbf) and pounds-mass (lbm).

WORK THROUGH EVERY PROBLEM

NCEES has greatly reduced the number of subjects about which you are expected to be knowledgeable and has made nothing optional. Skipping your weakest subjects is no longer a viable preparation strategy. You should study all examination knowledge areas, not just your specialty areas. That means you solve every problem in this book and skip nothing. Do not limit the number of problems you solve in hopes of finding enough problems in your areas of expertise to pass the exam.

The FE exam primarily uses SI units. Therefore, the need to work problems in both the customary U.S. and SI systems is greatly diminished. You will need to learn the SI system if you are not already familiar with it.

BE THOROUGH

Being thorough means really doing the work. Some people think they can read a problem statement, think about it for ten seconds, read the solution, and then say, "Yes, that's what I was thinking of, and that's what I would have done." Sadly, these people find out too late that the human brain doesn't learn very efficiently that way. Under pressure, they find they know and remember very little. For real learning, you'll have to spend some time with a stubby pencil.

There are so many places where you can get messed up solving a problem. Maybe it's in the use of your calculator, like pushing log instead of ln, or forgetting to set the angle to radians instead of degrees, and so on. Maybe it's rusty math. What is $\ln(e^x)$ anyway? How do you factor a polynomial? Maybe it's in finding the data needed or the proper unit conversion. Maybe you're not familiar with the SI system of units. These things take time. And you have to make the mistakes once so that you don't make them again.

If you do decide to get your hands dirty and actually work these problems, you'll have to decide how much reliance you place on this book. It's tempting to turn to a solution when you get slowed down by details or stumped by the subject material. It's tempting to want to maximize the number of problems you solve by spending as little time as possible solving them. However, you need to struggle a little bit more than that to really learn the material.

Studying a new subject is analogous to using a machete to cut a path through a dense jungle. By doing the work, you develop pathways that weren't there before. It's a lot different than just looking at the route on a map. You actually get nowhere by looking at a map. But cut the path once, and you're in business until the jungle overgrowth closes in again. So do the problems—all of them. And, don't look at the solutions until you've sweated a little.

1 Analytic Geometry and Trigonometry

PRACTICE PROBLEMS

1. To find the width of a river, a surveyor sets up a transit at point C on one river bank and sights directly across to point B on the other bank. The surveyor then walks along the bank for a distance of 275 m to point A. The angle CAB is $57° \, 28'$.

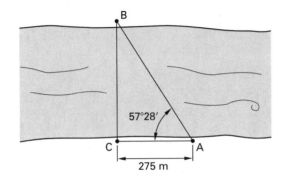

What is the approximate width of the river?

(A) 150 m

(B) 230 m

(C) 330 m

(D) 430 m

2. In the following illustration, angles 2 and 5 are $90°$, $AD = 15$, $DC = 20$, and $AC = 25$.

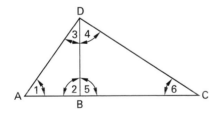

What are the lengths BC and BD, respectively?

(A) 12 and 16

(B) 13 and 17

(C) 16 and 12

(D) 18 and 13

3. What is the length of the line segment with slope 4/3 that extends from the point $(6, 4)$ to the y-axis?

(A) 10

(B) 25

(C) 50

(D) 75

4. Which of the following expressions is equivalent to $\sin 2\theta$?

(A) $2 \sin \theta \cos \theta$

(B) $\cos^2 \theta - \sin^2 \theta$

(C) $\sin \theta \cos \theta$

(D) $\dfrac{1 - \cos 2\theta}{2}$

5. Which of the following equations describes a circle with center at $(2, 3)$ and passing through the point $(-3, -4)$?

(A) $(x + 3)^2 + (y + 4)^2 = 85$

(B) $(x + 3)^2 + (y + 2)^2 = \sqrt{74}$

(C) $(x - 3)^2 + (y - 2)^2 = 74$

(D) $(x - 2)^2 + (y - 3)^2 = 74$

6. The equation for a circle is $x^2 + 4x + y^2 + 8y = 0$. What are the coordinates of the circle's center?

(A) $(-4, -8)$

(B) $(-4, -2)$

(C) $(-2, -4)$

(D) $(2, -4)$

7. Which of the following statements is FALSE for all noncircular ellipses?

(A) The eccentricity, e, is less than one.

(B) The ellipse has two foci.

(C) The sum of the two distances from the two foci to any point on the ellipse is $2a$ (i.e., twice the semimajor distance).

(D) The coefficients A and C preceding the x^2 and y^2 terms in the general form of the equation are equal.

8. What is the area of the shaded portion of the circle shown?

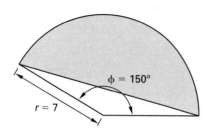

(A) $\dfrac{5\pi}{6} - 1$

(B) $\left(\dfrac{49}{12}\right)(5\pi - 3)$

(C) $\dfrac{50\pi}{3}$

(D) $49\pi - \sqrt{3}$

9. A pipe with a 20 cm inner diameter is filled to a depth equal to one-third of its diameter. What is the approximate area in flow?

(A) 33 cm^2

(B) 60 cm^2

(C) 92 cm^2

(D) 100 cm^2

10. The equation $y = a_1 + a_2 x$ is an algebraic expression for which of the following?

(A) a cosine expansion series

(B) projectile motion

(C) a circle in polar form

(D) a straight line

11. For the right triangle shown, $x = 18$ cm and $y = 13$ cm.

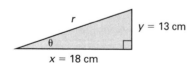

Most nearly, what is $\csc \theta$?

(A) 0.98

(B) 1.2

(C) 1.7

(D) 15

12. A circular sector has a radius of 8 cm and an arc length of 13 cm. Most nearly, what is its area?

(A) 48 cm^2

(B) 50 cm^2

(C) 52 cm^2

(D) 60 cm^2

13. The equation $-3x^2 - 4y^2 = 1$ defines

(A) a circle

(B) an ellipse

(C) a hyperbola

(D) a parabola

14. What is the approximate surface area (including both side and base) of a 4 m high right circular cone with a base 3 m in diameter?

(A) 24 m^2

(B) 27 m^2

(C) 32 m^2

(D) 36 m^2

15. A particle moves in the x-y plane. After t s, the x- and y-coordinates of the particle's location are $x = 8 \sin t$ and $y = 6 \cos t$. Which of the following equations describes the path of the particle?

(A) $36x^2 + 64y^2 = 2304$

(B) $36x^2 - 64y^2 = 2304$

(C) $64x^2 + 36y^2 = 2304$

(D) $64x^2 - 36y^2 = 2304$

SOLUTIONS

1. Use the formula for the tangent of an angle in a right triangle.

$$\tan\theta = BC/AC$$
$$BC = AC\tan\theta = (275\text{ m})\tan 57°\,28'$$
$$= 431.1\text{ m}\quad(430\text{ m})$$

The answer is (D).

2. For right triangle ABD,

$$(BD)^2 + (AB)^2 = (15)^2$$
$$(BD)^2 = (15)^2 - (AB)^2$$

For right triangle DBC,

$$(BD)^2 + (25 - AB)^2 = (20)^2$$
$$(BD)^2 = (20)^2 - (25 - AB)^2$$

Equate the two expressions for $(BD)^2$.

$$(15)^2 - (AB)^2 = (20)^2 - (25)^2 + 50(AB) - (AB)^2$$
$$AB = \frac{(15)^2 - (20)^2 + (25)^2}{50} = 9$$
$$BC = 25 - AB = 25 - 9 = 16$$
$$(BD)^2 = (15)^2 - (9)^2$$
$$BD = 12$$

Alternatively, this problem can be solved using the law of cosines.

The answer is (C).

3. The equation of the line is of the form

$$y = mx + b$$

The slope is $m = 4/3$, and a known point is $(x, y) = (6, 4)$. Find the y-intercept, b.

$$4 = \left(\tfrac{4}{3}\right)(6) + b$$
$$b = 4 - \left(\tfrac{4}{3}\right)(6) = -4$$

The complete equation is

$$y = \tfrac{4}{3}x - 4$$

b is the y-intercept, so the intersection with the y-axis is at point $(0, -4)$. The distance between these two points is

$$d = \sqrt{(y_2 - y_1)^2 + (x_2 - x_1)^2}$$
$$= \sqrt{(4 - (-4))^2 + (6 - 0)^2}$$
$$= 10$$

The answer is (A).

4. The double angle identity is

$$\sin 2\theta = 2\sin\theta\cos\theta$$

The answer is (A).

5. Substitute the known points into the center-radius form of the equation of a circle.

$$r^2 = (x - h)^2 + (y - k)^2$$
$$= (-3 - 2)^2 + (-4 - 3)^2$$
$$= 74$$

The equation of the circle is

$$(x - 2)^2 + (y - 3)^2 = 74$$

$(r^2 = 74$. The radius is $\sqrt{74}$.)

The answer is (D).

6. Use the standard form of the equation of a circle to find the circle's center.

$$x^2 + 4x + y^2 + 8y = 0$$
$$x^2 + 4x + 4 + y^2 + 8y + 16 = 4 + 16$$
$$(x + 2)^2 + (y + 4)^2 = 20$$

The center is at $(-2, -4)$.

The answer is (C).

7. The coefficients preceding the squared terms in the general equation are equal only for a straight line or circle, not for a noncircular ellipse.

$$Ax^2 + Bxy + Cy^2 + Dx + Ey + F = 0$$

The answer is (D).

8. The area of the circle is

$$\phi = (150°)\left(\frac{2\pi\text{ rad}}{360°}\right) = \frac{5\pi}{6}\text{ rad}$$
$$A = \frac{r^2(\phi - \sin\phi)}{2}$$
$$= \frac{(7)^2\left(\dfrac{5\pi}{6} - \sin\dfrac{5\pi}{6}\right)}{2}$$
$$= \left(\frac{49}{2}\right)\left(\frac{5\pi}{6} - \frac{1}{2}\right)$$
$$= \left(\frac{49}{12}\right)(5\pi - 3)$$

The answer is (B).

9. Find the angle ϕ.

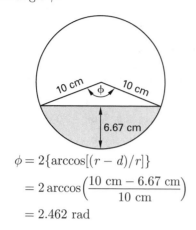

$$\phi = 2\{\arccos[(r-d)/r]\}$$
$$= 2\arccos\left(\frac{10 \text{ cm} - 6.67 \text{ cm}}{10 \text{ cm}}\right)$$
$$= 2.462 \text{ rad}$$

Find the area of flow.

$$A = [r^2(\phi - \sin\phi)]/2$$
$$= \frac{(10 \text{ cm})^2\left(2.46 \text{ rad} - \sin(2.462 \text{ rad})\right)}{2}$$
$$= 91.67 \text{ cm}^2 \quad (92 \text{ cm}^2)$$

The answer is (C).

10. $y = mx + b$ is the slope-intercept form of the equation of a straight line. a_1 and a_2 are both constants, so $y = a_1 + a_2 x$ describes a straight line.

The answer is (D).

11. Find the length of the hypotenuse, r.

$$r = \sqrt{x^2 + y^2} = \sqrt{(18 \text{ cm})^2 + (13 \text{ cm})^2} = 22.2 \text{ cm}$$

Find $\csc\theta$.

$$\csc\theta = r/y = \frac{22.2 \text{ cm}}{13 \text{ cm}} = 1.7$$

The answer is (C).

12. Find the area of the circular sector.

$$A = sr/2 = \frac{(13 \text{ cm})(8 \text{ cm})}{2} = 52 \text{ cm}^2$$

The answer is (C).

13. The general form of the conic section equation is

$$Ax^2 + Bxy + Cy^2 + Dx + Ey + F = 0$$

$A = -3$, $C = -4$, $F = -1$, and $B = D = E = 0$. A and C are different, so the equation does not define a circle. Calculate the discriminant.

$$B^2 - 4AC = (0)^2 - (4)(-3)(-4) = -48$$

This is less than zero, so the equation defines an ellipse.

The answer is (B).

14. Find the total surface area of a right circular cone. The radius is $r = d/2 = 3 \text{ m}/2 = 1.5 \text{ m}$.

$$A = \text{side area} + \text{base area} = \pi r\left(r + \sqrt{r^2 + h^2}\right)$$
$$= \pi(1.5 \text{ m})\left(1.5 \text{ m} + \sqrt{(1.5 \text{ m})^2 + (4 \text{ m})^2}\right)$$
$$= 27.2 \text{ m}^2 \quad (27 \text{ m}^2)$$

The answer is (B).

15. Rearrange the two coordinate equations.

$$\sin t = \frac{x}{8}$$
$$\cos t = \frac{y}{6}$$

Use the following trigonometric identity.

$$\sin^2\theta + \cos^2\theta = 1$$
$$\left(\frac{x}{8}\right)^2 + \left(\frac{y}{6}\right)^2 = 1$$

To clear the fractions, multiply both sides by $(8)^2 \times (6)^2 = 2304$.

$$36x^2 + 64y^2 = 2304$$

The answer is (A).

 **Algebra and Linear Algebra**

PRACTICE PROBLEMS

1. What is the name for a vector that represents the sum of two vectors?

(A) scalar

(B) resultant

(C) tensor

(D) moment

2. The second and sixth terms of a geometric progression are 3/10 and 243/160, respectively. What is the first term of this sequence?

(A) 1/10

(B) 1/5

(C) 3/5

(D) 3/2

3. Using logarithmic identities, what is most nearly the numerical value for the following expression?

$$\log_3 \tfrac{3}{2} + \log_3 12 - \log_3 2$$

(A) 0.95

(B) 1.33

(C) 2.00

(D) 2.20

4. Which of the following statements is true for a power series with the general term $a_i x^i$?

I.   An infinite power series converges for $x < 1$.

II. Power series can be added together or subtracted within their interval of convergence.

III. Power series can be integrated within their interval of convergence.

(A) I only

(B) II only

(C) I and III

(D) II and III

5. What is most nearly the length of the resultant of the following vectors?

$$3\mathbf{i} + 4\mathbf{j} - 5\mathbf{k}$$
$$7\mathbf{i} + 2\mathbf{j} + 3\mathbf{k}$$
$$-16\mathbf{i} - 14\mathbf{j} + 2\mathbf{k}$$

(A) 3

(B) 4

(C) 10

(D) 14

6. What is the solution to the following system of simultaneous linear equations?

$$10x + 3y + 10z = 5$$
$$8x - 2y + 9z = 3$$
$$8x + y - 10z = 7$$

(A) $x = 0.326$; $y = -0.192$; $z = 0.586$

(B) $x = 0.148$; $y = 1.203$; $z = 0.099$

(C) $x = 0.625$; $y = 0.186$; $z = -0.181$

(D) $x = 0.282$; $y = -1.337$; $z = -0.131$

7. What is the inverse of matrix $\mathbf{A}$?

$$\mathbf{A} = \begin{bmatrix} 2 & 3 \\ 1 & 1 \end{bmatrix}$$

(A) $\begin{bmatrix} 2 & 3 \\ 1 & 1 \end{bmatrix}$

(B) $\begin{bmatrix} 3 & 2 \\ 1 & 1 \end{bmatrix}$

(C) $\begin{bmatrix} 1 & -3 \\ -1 & 2 \end{bmatrix}$

(D) $\begin{bmatrix} -1 & 3 \\ 1 & -2 \end{bmatrix}$

8. If the determinant of matrix **A** is -40, what is the determinant of matrix **B**?

$$\mathbf{A} = \begin{bmatrix} 4 & 3 & 2 & 1 \\ 0 & 1 & 2 & -1 \\ 2 & 3 & -1 & 1 \\ 1 & 1 & 1 & 2 \end{bmatrix} \qquad \mathbf{B} = \begin{bmatrix} 2 & 1.5 & 1 & 0.5 \\ 0 & 1 & 2 & -1 \\ 2 & 3 & -1 & 1 \\ 1 & 1 & 1 & 2 \end{bmatrix}$$

(A) -80

(B) -40

(C) -20

(D) 0.5

9. Given the origin-based vector $\mathbf{A} = \mathbf{i} + 2\mathbf{j} + \mathbf{k}$, what is most nearly the angle between **A** and the x-axis?

(A) $22°$

(B) $24°$

(C) $66°$

(D) $80°$

10. Which is a true statement about these two vectors?

$$\mathbf{A} = \mathbf{i} + 2\mathbf{j} + \mathbf{k}$$
$$\mathbf{B} = \mathbf{i} + 3\mathbf{j} - 7\mathbf{k}$$

(A) Both vectors pass through the point $(0, -1, 6)$.

(B) The vectors are parallel.

(C) The vectors are orthogonal.

(D) The angle between the vectors is $17.4°$.

11. What is most nearly the acute angle between vectors $\mathbf{A} = (3, 2, 1)$ and $\mathbf{B} = (2, 3, 2)$, both based at the origin?

(A) $25°$

(B) $33°$

(C) $35°$

(D) $59°$

12. Force vectors **A**, **B**, and **C** are applied at a single point.

$$\mathbf{A} = \mathbf{i} + 3\mathbf{j} + 4\mathbf{k}$$
$$\mathbf{B} = 2\mathbf{i} + 7\mathbf{j} - \mathbf{k}$$
$$\mathbf{C} = -\mathbf{i} + 4\mathbf{j} + 2\mathbf{k}$$

What is most nearly the magnitude of the resultant force vector, **R**?

(A) 13

(B) 14

(C) 15

(D) 16

13. What is the sum of $12 + 13j$ and $7 - 9j$?

(A) $19 - 22j$

(B) $19 + 4j$

(C) $25 - 22j$

(D) $25 + 4j$

14. What is the product of the complex numbers $3 + 4j$ and $7 - 2j$?

(A) $10 + 2j$

(B) $13 + 22j$

(C) $13 + 34j$

(D) $29 + 22j$

SOLUTIONS

1. By definition, the sum of two vectors is known as the resultant.

The answer is (B).

2. The common ratio is

$$l = ar^{n-1}$$

$$\frac{l_6}{l_2} = \frac{ar^{6-1}}{ar^{2-1}} = r^4$$

$$r = \sqrt[4]{\frac{l_6}{l_2}}$$

$$= \sqrt[4]{\frac{\dfrac{243}{160}}{\dfrac{3}{10}}}$$

$$= 3/2$$

The term before $3/10$ is

$$a_1 = \frac{\dfrac{3}{10}}{\dfrac{3}{2}} = 1/5$$

The answer is (B).

3. Use the following logarithmic identities.

$$\log xy = \log x + \log y$$

$$\log x/y = \log x - \log y$$

$$\log_3 \frac{3}{2} + \log_3 12 - \log_3 2 = \log_3 \frac{\left(\dfrac{3}{2}\right)(12)}{2}$$

$$= \log_3 9$$

Since $(3)^2 = 9$,

$$\log_3 9 = 2.00$$

The answer is (C).

4. Power series can be added together, subtracted from each other, differentiated, and integrated within their interval of convergence. The interval of convergence is $-1 < x < 1$.

The answer is (D).

5. The resultant is produced by adding the vectors.

$$
\begin{array}{r}
3\mathbf{i} + 4\mathbf{j} - 5\mathbf{k} \\
7\mathbf{i} + 2\mathbf{j} + 3\mathbf{k} \\
-16\mathbf{i} - 14\mathbf{j} + 2\mathbf{k} \\
\hline
-6\mathbf{i} - 8\mathbf{j} + 0\mathbf{k}
\end{array}
$$

The length of the resultant vector is

$$|\mathbf{R}| = \sqrt{(-6)^2 + (-8)^2 + (0)^2}$$

$$= 10$$

The answer is (C).

6. There are several ways of solving this problem.

$$\mathbf{AX} = \mathbf{B}$$

$$\begin{bmatrix} 10 & 3 & 10 \\ 8 & -2 & 9 \\ 8 & 1 & -10 \end{bmatrix} \begin{bmatrix} x \\ y \\ z \end{bmatrix} = \begin{bmatrix} 5 \\ 3 \\ 7 \end{bmatrix}$$

$$\mathbf{AA}^{-1}\mathbf{X} = \mathbf{A}^{-1}\mathbf{B}$$

$$\mathbf{IX} = \mathbf{A}^{-1}\mathbf{B}$$

$$\mathbf{X} = \mathbf{A}^{-1}\mathbf{B}$$

$$\mathbf{X} = \begin{bmatrix} \dfrac{11}{806} & \dfrac{20}{403} & \dfrac{47}{806} \\ \dfrac{76}{403} & \dfrac{-90}{403} & \dfrac{-5}{403} \\ \dfrac{12}{403} & \dfrac{7}{403} & \dfrac{-22}{403} \end{bmatrix} \begin{bmatrix} 5 \\ 3 \\ 7 \end{bmatrix}$$

$$= \begin{bmatrix} (5)\left(\dfrac{11}{806}\right) + (3)\left(\dfrac{20}{403}\right) + (7)\left(\dfrac{47}{806}\right) \\ (5)\left(\dfrac{76}{403}\right) + (3)\left(\dfrac{-90}{403}\right) + (7)\left(\dfrac{-5}{403}\right) \\ (5)\left(\dfrac{12}{403}\right) + (3)\left(\dfrac{7}{403}\right) + (7)\left(\dfrac{-22}{403}\right) \end{bmatrix}$$

$$= \begin{bmatrix} 0.625 \\ 0.186 \\ -0.181 \end{bmatrix}$$

(Direct substitution of the four answer choices into the original equations is probably the fastest way of solving this type of problem.)

The answer is (C).

7. Find the determinant.

$$|\mathbf{A}| = 2 \times 1 - 1 \times 3 = -1$$

The inverse of a 2×2 matrix is

$$\mathbf{A}^{-1} = \frac{\text{adj}(\mathbf{A})}{|\mathbf{A}|} = \frac{\begin{bmatrix} b_2 & -a_2 \\ -b_1 & a_1 \end{bmatrix}}{|\mathbf{A}|}$$

$$= \frac{\begin{bmatrix} 1 & -3 \\ -1 & 2 \end{bmatrix}}{-1}$$

$$= \begin{bmatrix} -1 & 3 \\ 1 & -2 \end{bmatrix}$$

The answer is (D).

8. The first row of matrix $\mathbf{B}$ is half that of $\mathbf{A}$, and the other rows are the same in $\mathbf{A}$ and $\mathbf{B}$, so the determinant of $\mathbf{B}$ is half the determinant of $\mathbf{A}$.

The answer is (C).

9. The magnitude of vector $\mathbf{A}$ is

$$|\mathbf{A}| = \sqrt{(1)^2 + (2)^2 + (1)^2} = \sqrt{6}$$

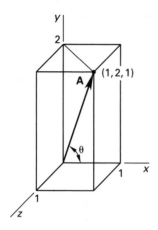

The x-component of the vector is 1, so the direction cosine is

$$\cos\theta_x = \frac{1}{\sqrt{6}}$$

The angle is

$$\theta = \cos^{-1}\left(\frac{1}{\sqrt{6}}\right) = 65.9° \quad (66°)$$

The answer is (C).

10. The magnitudes of the two vectors are

$$|\mathbf{A}| = \sqrt{(1)^2 + (2)^2 + (1)^2} = \sqrt{6}$$

$$|\mathbf{B}| = \sqrt{(1)^2 + (3)^2 + (-7)^2} = \sqrt{59}$$

The angle between them is

$$\phi = \cos^{-1}\left(\frac{a_x b_x + a_y b_y + a_z b_z}{|\mathbf{A}||\mathbf{B}|}\right)$$

$$= \cos^{-1}\left(\frac{(1)(1) + (2)(3) + (1)(-7)}{\sqrt{6}\sqrt{59}}\right)$$

$$= 90°$$

The vectors are orthogonal.

The answer is (C).

11. The angle between the two vectors is

$$\theta = \cos^{-1}\left(\frac{\mathbf{A}\cdot\mathbf{B}}{|\mathbf{A}||\mathbf{B}|}\right)$$

$$= \cos^{-1}\left(\frac{a_x b_x + a_y b_y + a_z b_z}{|\mathbf{A}||\mathbf{B}|}\right)$$

$$= \cos^{-1}\left(\frac{(3)(2) + (2)(3) + (1)(2)}{\sqrt{(3)^2 + (2)^2 + (1)^2}\sqrt{(2)^2 + (3)^2 + (2)^2}}\right)$$

$$= 24.8° \quad (25°)$$

The answer is (A).

12. The magnitude of $\mathbf{R}$ is

$$|\mathbf{R}| = \sqrt{(1 + 2 - 1)^2 + (3 + 7 + 4)^2 + (4 - 1 + 2)^2}$$

$$= \sqrt{4 + 196 + 25}$$

$$= \sqrt{225}$$

$$= 15$$

The answer is (C).

13. Add the real parts and the imaginary parts of each complex number.

$$(a + jb) + (c + jd) = (a + c) + j(b + d)$$

$$(12 + 13j) + (7 - 9j) = (12 + 7) + j(13 + (-9))$$

$$= 19 + 4j$$

The answer is (B).

14. Use the algebraic distributive law and the equivalency $j^2 = -1$.

$$(a + jb)(c + jd) = (ac - bd) + j(ad + bc)$$

$$(3 + 4j)(7 - 2j) = 21 - 8j^2 + 28j - 6j$$

$$= 21 + 8 + 28j - 6j$$

$$= 29 + 22j$$

The answer is (D).

3 Calculus

PRACTICE PROBLEMS

1. Which of the following is NOT a correct derivative?

(A) $\dfrac{d}{dx}\cos x = -\sin x$

(B) $\dfrac{d}{dx}(1-x)^3 = -3(1-x)^2$

(C) $\dfrac{d}{dx}\dfrac{1}{x} = -\dfrac{1}{x^2}$

(D) $\dfrac{d}{dx}\csc x = -\cot x$

2. What is the derivative, dy/dx, of the expression $x^2y - e^{2x} = \sin y$?

(A) $\dfrac{2e^{2x}}{x^2 - \cos y}$

(B) $\dfrac{2e^{2x} - 2xy}{x^2 - \cos y}$

(C) $2e^{2x} - 2xy$

(D) $x^2 - \cos y$

3. What is the approximate area bounded by the curves $y = 8 - x^2$ and $y = -2 + x^2$?

(A) 22

(B) 27

(C) 30

(D) 45

4. What are the minimum and maximum values, respectively, of the equation $f(x) = 5x^3 - 2x^2 + 1$ on the interval $[-2, 2]$?

(A) $-47, 33$

(B) $-4, 4$

(C) $0.95, 1$

(D) $0, 0.27$

5. In vector calculus, a gradient is a

I. vector that points in the direction of the rate of change of a scalar field

II. vector whose magnitude indicates the maximum rate of change of a scalar field

III. scalar that indicates the magnitude of the rate of change of a vector field in a particular direction

IV. scalar that indicates the maximum magnitude of the rate of change of a vector field in any particular direction

(A) I only

(B) III only

(C) I and II

(D) III and IV

6. Which of the illustrations shown represents the vector field, $\mathbf{F}(x, y) = -y\mathbf{i} + x\mathbf{j}$?

(A)

(B)

(C)

(D)

7. A peach grower estimates that if he picks his crop now, he will obtain 1000 lugs of peaches, which he can sell at \$1.00 per lug. However, he estimates that his crop will increase by an additional 60 lugs of peaches for each week that he delays picking, but the price will drop at a rate of \$0.025 per lug per week. In addition, he will experience a spoilage rate of approximately 10 lugs for each week he delays. In order to maximize his revenue, how many weeks should he wait before picking the peaches?

(A) 2 weeks

(B) 5 weeks

(C) 7 weeks

(D) 10 weeks

8. Determine the following indefinite integral.

$$\int \frac{x^3 + x + 4}{x^2}\, dx$$

(A) $\dfrac{x}{4} + \ln|x| - \dfrac{4}{x} + C$

(B) $\dfrac{-x}{2} + \log x - 8x + C$

(C) $\dfrac{x^2}{2} + \ln|x| - \dfrac{2}{x^2} + C$

(D) $\dfrac{x^2}{2} + \ln|x| - \dfrac{4}{x} + C$

9. Find dy/dx for the parametric equations given.

$$x = 2t^2 - t$$
$$y = t^3 - 2t + 1$$

(A) $3t^2$

(B) $3t^2/2$

(C) $4t - 1$

(D) $(3t^2 - 2)/(4t - 1)$

10. A two-dimensional function, $f(x, y)$, is defined as

$$f(x, y) = 2x^2 - y^2 + 3x - y$$

What is the direction of the line passing through the point $(1, -2)$ that has the maximum slope?

(A) $4\mathbf{i} + 2\mathbf{j}$

(B) $7\mathbf{i} + 3\mathbf{j}$

(C) $7\mathbf{i} + 4\mathbf{j}$

(D) $9\mathbf{i} - 7\mathbf{j}$

11. Evaluate the following limit.

$$\lim_{x \to 2} \left(\frac{x^2 - 4}{x - 2} \right)$$

(A) 0

(B) 2

(C) 4

(D) ∞

12. If $f(x, y) = x^2y^3 + xy^4 + \sin x + \cos^2 x + \sin^3 y$, what is $\partial f/\partial x$?

(A) $(2x + y)y^3 + 3\sin^2 y \cos y$

(B) $(4x - 3y^2)xy^2 + 3\sin^2 y \cos y$

(C) $(3x + 4y^2)xy + 3\sin^2 y \cos y$

(D) $(2x + y)y^3 + (1 - 2\sin x)\cos x$

13. What is dy/dx if $y = (2x)^x$?

(A) $(2x)^x(2 + \ln 2x)$

(B) $2x(1 + \ln 2x)^x$

(C) $(2x)^x(\ln 2x^2)$

(D) $(2x)^x(1 + \ln 2x)$

SOLUTIONS

1. Determine each of the derivatives.

$$\frac{d}{dx}\cos x = -\sin x \quad \text{[OK]}$$

$$\frac{d}{dx}(1-x)^3 = (3)(1-x)^2(-1) = (-3)(1-x)^2 \quad \text{[OK]}$$

$$\frac{d}{dx}\frac{1}{x} = \frac{d}{dx}\,x^{-1} = (-1)(x^{-2}) = \frac{-1}{x^2} \quad \text{[OK]}$$

$$\frac{d}{dx}\csc x = -\cot x \quad \text{[incorrect]}$$

The answer is (D).

2. Since neither x nor y can be extracted from the equality, rearrange to obtain a homogeneous expression in x and y.

$$x^2 y - e^{2x} = \sin y$$

$$f(x,y) = x^2 y - e^{2x} - \sin y = 0$$

Take the partial derivatives with respect to x and y.

$$\frac{\partial f(x,y)}{\partial x} = 2xy - 2e^{2x}$$

$$\frac{\partial f(x,y)}{\partial y} = x^2 - \cos y$$

Use implicit differentiation.

$$\frac{\partial y}{\partial x} = \frac{-\dfrac{\partial f(x,y)}{\partial x}}{\dfrac{\partial f(x,y)}{\partial y}} = \frac{2e^{2x} - 2xy}{x^2 - \cos y}$$

The answer is (B).

3. Find the intersection points by setting the two functions equal.

$$-2 + x^2 = 8 - x^2$$

$$2x^2 = 10$$

$$x = \pm\sqrt{5}$$

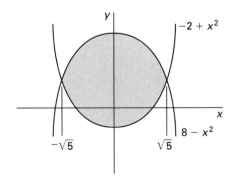

The integral of $f_1(x) - f_2(x)$ represents the area between the two curves between the limits of integration.

$$A = \int_{x_1}^{x_2}\left(f_1(x) - f_2(x)\right)dx$$

$$= \int_{-\sqrt{5}}^{\sqrt{5}}\left((8 - x^2) - (-2 + x^2)\right)dx$$

$$= \int_{-\sqrt{5}}^{\sqrt{5}}(10 - 2x^2)\,dx$$

$$= \left(10x - \tfrac{2}{3}x^3\right)\Big|_{-\sqrt{5}}^{\sqrt{5}}$$

$$= 29.8 \quad (30)$$

The answer is (C).

4. The critical points are located where the first derivative is zero.

$$f(x) = 5x^3 - 2x^2 + 1$$

$$f'(x) = 15x^2 - 4x$$

$$15x^2 - 4x = 0$$

$$x(15x - 4) = 0$$

$$x = 0 \quad \text{or} \quad x = 4/15$$

Test for a maximum, minimum, or inflection point.

$$f''(x) = 30x - 4$$

$$f''(0) = (30)(0) - 4$$

$$= -4$$

$$f''(a) < 0 \quad \text{[maximum]}$$

$$f''\left(\frac{4}{15}\right) = (30)\left(\frac{4}{15}\right) - 4$$

$$= 4$$

$$f''(a) > 0 \quad \text{[minimum]}$$

These could be a local maximum and minimum. Check the endpoints of the interval and compare with the function values at the critical points.

$$f(-2) = (5)(-2)^3 - (2)(-2)^2 + 1 = -47$$

$$f(2) = (5)(2)^3 - (2)(2)^2 + 1 = 33$$

$$f(0) = (5)(0)^3 - (2)(0)^2 + 1 = 1$$

$$f\left(\frac{4}{15}\right) = (5)\left(\frac{4}{15}\right)^3 - (2)\left(\frac{4}{15}\right)^2 + 1$$

$$= 0.95$$

The minimum and maximum values of the equation, -47 and 33, respectively, are at the endpoints.

The answer is (A).

5. A gradient, such as the slope of a straight line, is a pure number. It is not a vector, and it indicates no direction. However, it is understood that the rate of change of the dependent variable is always with respect to increases in the independent variables. In that sense, the gradient indicates the rate of change of the dependent variable in a particular direction. Except for straight lines, the gradient depends on the values of the independent variables (i.e., on the location). A value for a particular location may or may not be the maximum possible value over the entire range of independent variable values.

The answer is (B).

6. From $-y\mathbf{i}$, it can be concluded that for

(a) positive values of y, the vector field points to the left, and

(b) negative values of y, the vector field points to the right.

From $+x\mathbf{j}$, it can be concluded that for

(a) positive values of x, the vector field points upward, and

(b) negative values of x, the vector field points downward.

The answer is (C).

7. Let x represent the number of weeks.

The equation describing the price as a function of time is

$$\frac{\text{price}}{\text{lug}} = \$1 - \$0.025x$$

The equation describing the yield is

$$\text{lugs sold} = 1000 + (60 - 10)x$$
$$= 1000 + 50x$$

The revenue function is

$$R = \left(\frac{\text{price}}{\text{lug}}\right)(\text{lugs sold})$$
$$= (1 - 0.025x)(1000 + 50x)$$
$$= 1000 + 50x - 25x - 1.25x^2$$
$$= 1000 + 25x - 1.25x^2$$

To maximize the revenue function, set its derivative equal to zero.

$$\frac{dR}{dx} = 25 - 2.5x = 0$$
$$x = 10 \text{ weeks}$$

The answer is (D).

8. Separate the fraction into parts and integrate each one.

$$\int \frac{x^3 + x + 4}{x^2}\,dx = \int \frac{x^3}{x^2}\,dx + \int \frac{x}{x^2}\,dx + \int \frac{4}{x^2}\,dx$$
$$= \int x\,dx + \int \frac{1}{x}\,dx + 4\int \frac{1}{x^2}\,dx$$
$$= \frac{x^2}{2} + \ln|x| + 4\left(\frac{x^{-1}}{-1}\right) + C$$
$$= \frac{x^2}{2} + \ln|x| - \frac{4}{x} + C$$

The answer is (D).

9. Calculate the derivatives of x and y with respect to t.

$$\frac{dy}{dt} = 3t^2 - 2$$
$$\frac{dx}{dt} = 4t - 1$$

The derivative of y with respect to x is

$$\frac{dy}{dx} = \frac{\dfrac{dy}{dt}}{\dfrac{dx}{dt}}$$
$$= \frac{3t^2 - 2}{4t - 1}$$

The answer is (D).

10. The direction of the line passing through $(1, -2)$ with maximum slope is found by inserting $x = 1$ and $y = -2$ into the gradient vector function.

The gradient of the function is

$$\nabla f(x, y, z) = \frac{\partial f(x, y, z)}{\partial x}\mathbf{i} + \frac{\partial f(x, y, z)}{\partial y}\mathbf{j} + \frac{\partial f(x, y, z)}{\partial z}\mathbf{k}$$
$$= \frac{\partial(2x^2 - y^2 + 3x - y)}{\partial x}\mathbf{i}$$
$$+ \frac{\partial(2x^2 - y^2 + 3x - y)}{\partial y}\mathbf{j}$$
$$= (4x + 3)\mathbf{i} - (2y + 1)\mathbf{j}$$

At $(1, -2)$,

$$\nabla f(1, -2) = \big((4)(1) + 3\big)\mathbf{i} - \big((2)(-2) + 1\big)\mathbf{j}$$
$$= 7\mathbf{i} + 3\mathbf{j}$$

The answer is (B).

11. The expression approaches $0/0$ at the limit.

$$\frac{(2)^2 - 4}{2 - 2} = \frac{0}{0}$$

Use L'Hôpital's rule.

$$\lim_{x \to 2}\left(\frac{x^2 - 4}{x - 2}\right) = \lim_{x \to 2}\left(\frac{\dfrac{d}{dx}(x^2 - 4)}{\dfrac{d}{dx}(x - 2)}\right) = \lim_{x \to 2}\left(\frac{2x}{1}\right)$$

$$= \frac{(2)(2)}{1}$$

$$= 4$$

This could also be solved by factoring the numerator.

The answer is (C).

12. The partial derivative with respect to x is found by treating all other variables as constants. Therefore, all terms that do not contain x have zero derivatives.

$$\frac{\partial f}{\partial x} = 2xy^3 + y^4 + \cos x + 2\cos x(-\sin x)$$

$$= (2x + y)y^3 + (1 - 2\sin x)\cos x$$

The answer is (D).

13. From the table of derivatives,

$$\mathbf{D}\big(f(x)\big)^{g(x)} = g(x)\big(f(x)\big)^{g(x)-1}\mathbf{D}f(x)$$
$$+ \ln\big(f(x)\big)\big(f(x)\big)^{g(x)}\mathbf{D}g(x)$$

$$f(x) = 2x$$

$$g(x) = x$$

$$\frac{d(2x)^x}{dx} = x(2x)^{x-1}(2) + (\ln 2x)(2x)^x(1)$$

$$= (2x)^x + (2x)^x \ln 2x$$

$$= (2x)^x(1 + \ln 2x)$$

The answer is (D).

4 Differential Equations and Transforms

PRACTICE PROBLEMS

1. What is the solution to the following differential equation?

$$y' + 5y = 0$$

(A) $y = 5x + C$

(B) $y = Ce^{-5x}$

(C) $y = Ce^{5x}$

(D) either (A) or (B)

2. What is the solution to the following linear difference equation?

$$(k+1)\big(y(k+1)\big) - ky(k) = 1$$

(A) $y(k) = 12 - \dfrac{1}{k}$

(B) $y(k) = 1 - \dfrac{12}{k}$

(C) $y(k) = 12 + 3k$

(D) $y(k) = 3 + \dfrac{1}{k}$

3. What is the general solution to the following differential equation?

$$2\left(\frac{d^2 y}{dx^2}\right) - 4\left(\frac{dy}{dx}\right) + 4y = 0$$

(A) $y = C_1 \cos x + C_2 \sin x$

(B) $y = C_1 e^x + C_2 e^{-x}$

(C) $y = e^{-x}(C_1 \cos x - C_2 \sin x)$

(D) $y = e^x(C_1 \cos x + C_2 \sin x)$

4. What is the general solution to the following differential equation?

$$\frac{d^2 y}{dx^2} + 2\frac{dy}{dx} + 2y = 0$$

(A) $y = C_1 \sin x - C_2 \cos x$

(B) $y = C_1 \cos x - C_2 \sin x$

(C) $y = C_1 \cos x + C_2 \sin x$

(D) $y = e^{-x}(C_1 \cos x + C_2 \sin x)$

5. What is the complementary solution to the following differential equation?

$$y'' - 4y' + \tfrac{25}{4}y = 10\cos 8x$$

(A) $y = 2C_1 x + C_2 x - C_3 x$

(B) $y = C_1 e^{2x} + C_2 e^{1.5x}$

(C) $y = C_1 e^{2x} \cos 1.5x + C_2 e^{2x} \sin 1.5x$

(D) $y = C_1 e^x \tan x + C_2 e^x \cot x$

6. What is the general solution to the following differential equation?

$$y'' + y' + y = 0$$

(A) $y = e^{-\frac{1}{2}x}\left(C_1 \cos \frac{\sqrt{3}}{2}x + C_2 \sin \frac{\sqrt{3}}{2}x\right)$

(B) $y = e^{-\frac{1}{2}x}\left(C_1 \cos \frac{3}{2}x + C_2 \sin \frac{3}{2}x\right)$

(C) $y = e^{-2x}\left(C_1 \cos \frac{\sqrt{3}}{2}x + C_2 \sin \frac{\sqrt{3}}{2}x\right)$

(D) $y = e^{-2x}\left(C_1 \cos \frac{3}{2}x + C_2 \sin \frac{3}{2}x\right)$

7. What is the solution to the following differential equation if $x = 1$ at $t = 0$, and $dx/dt = 0$ at $t = 0$?

$$\tfrac{1}{2}\frac{d^2 x}{dt^2} + 4\frac{dx}{dt} + 8x = 5$$

(A) $x = e^{-4t} + 4te^{-4t}$

(B) $x = \frac{3}{8}e^{-2t}(\cos 2t + \sin 2t) + \frac{5}{8}$

(C) $x = e^{-4t} + 4te^{-4t} + \frac{5}{8}$

(D) $x = \frac{3}{8}e^{-4t} + \frac{3}{2}te^{-4t} + \frac{5}{8}$

8. In the following differential equation with the initial condition $x(0) = 12$, what is the value of $x(2)$?

$$\frac{dx}{dt} + 4x = 0$$

(A) 3.4×10^{-3}

(B) 4.0×10^{-3}

(C) 5.1×10^{-3}

(D) 6.2×10^{-3}

9. What are the three general Fourier coefficients for the sawtooth wave shown?

(A) $a_0 = 0$, $a_n = 0$, $b_n = \dfrac{-1}{\pi n}$

(B) $a_0 = \dfrac{1}{2}$, $a_n = 0$, $b_n = \dfrac{-1}{\pi n}$

(C) $a_0 = 1$, $a_n = 1$, $b_n = \dfrac{1}{\pi n}$

(D) $a_0 = \dfrac{1}{2}$, $a_n = \dfrac{1}{2}$, $b_n = \dfrac{1}{\pi n}$

10. The values of an unknown function follow a Fibonacci number sequence. It is known that $f(1) = 4$ and $f(2) = 1.3$. What is $f(4)$?

(A) -4.1

(B) 0.33

(C) 2.7

(D) 6.6

SOLUTIONS

1. This is a first-order linear equation with characteristic equation $r + 5 = 0$. The form of the solution is

$$y = Ce^{-5x}$$

In the preceding equation, the constant, C, could be determined from additional information.

The answer is (B).

2. Since nothing is known about the general form of $y(k)$, the only way to solve this problem is by trial and error, substituting each answer option into the equation in turn. Option B is

$$y(k) = 1 - \frac{12}{k}$$

Substitute this into the difference equation.

$$(k+1)\big(y(k+1)\big) - k\big(y(k)\big) = 1$$

$$(k+1)\left(1 - \frac{12}{k+1}\right) - k\left(1 - \frac{12}{k}\right) = 1$$

$$(k+1)\left(\frac{k+1-12}{k+1}\right) - k\left(\frac{k-12}{k}\right) = 1$$

$$k + 1 - 12 - k + 12 = 1$$

$$1 = 1$$

$y(k) = 1 - 12/k$ solves the difference equation.

The answer is (B).

3. This is a second-order, homogeneous, linear differential equation. Start by putting it in general form.

$$y'' + 2ay' + by = 0$$

$$2y'' - 4y' + 4y = 0$$

$$y'' - 2y' + 2y = 0$$

$$a = -2$$

$$b = 2$$

Since $a^2 < 4b$, the form of the equation is

$$y = e^{\alpha x}(C_1 \cos \beta x + C_2 \sin \beta_x)$$

$$\alpha = \frac{-a}{2} = \frac{-2}{2} = 1$$

$$\beta = \frac{\sqrt{4b - a^2}}{2}$$

$$= \frac{\sqrt{(4)(2) - (-2)^2}}{2}$$

$$= 1$$

$$y = e^x(C_1 \cos x + C_2 \sin x)$$

The answer is (D).

4. The characteristic equation is

$$r^2 + 2r + 2 = 0$$
$$a = 2$$
$$b = 2$$

The roots are

$$r_{1,2} = \frac{-a \pm \sqrt{a^2 - 4b}}{2}$$
$$= \frac{-2 \pm \sqrt{(2)^2 - (4)(2)}}{2}$$
$$= (-1 + i), (-1 - i)$$

Since $a^2 < 4b$, the solution is

$$y = e^{\alpha x}(C_1 \cos \beta x + C_2 \sin \beta x)$$
$$\alpha = \frac{-a}{2} = \frac{-2}{2} = -1$$
$$\beta = \frac{\sqrt{4b - a^2}}{2} = \frac{\sqrt{(4)(2) - (2)^2}}{2}$$
$$= 1$$
$$y = e^{-x}(C_1 \cos x + C_2 \sin x)$$

The answer is (D).

5. The complementary solution to a nonhomogeneous differential equation is the solution of the homogeneous differential equation.

The characteristic equation is

$$r^2 + ar + b = 0$$
$$r^2 - 4r + \frac{25}{4} = 0$$

So, $a = -4$, and $b = 25/4$.

The roots are

$$r_{1,2} = \frac{-a \pm \sqrt{a^2 - 4b}}{2}$$
$$= \frac{-(-4) \pm \sqrt{(-4)^2 - (4)\left(\frac{25}{4}\right)}}{2}$$
$$= 2 \pm 1.5i$$

Since the roots are imaginary, the homogeneous solution has the form of

$$y = e^{\alpha x}(C_1 \cos \beta x + C_2 \sin \beta x)$$
$$\alpha = 2$$
$$\beta = \pm 1.5$$

The complementary solution is

$$y = e^{2x}(C_1 \cos 1.5x + C_2 \sin 1.5x)$$
$$= C_1 e^{2x} \cos 1.5x + C_2 e^{2x} \sin 1.5x$$

The answer is (C).

6. This is a second-order, homogeneous, linear differential equation with $a = b = 1$. This differential equation can be solved by the method of undetermined coefficients with a solution in the form $y = Ce^{rx}$. The substitution of the solution gives

$$(r^2 + ar + b)Ce^{rx} = 0$$

Because Ce^{rx} can never be zero, the characteristic equation is

$$r^2 + ar + b = 0$$

Because $a^2 = 1 < 4b = 4$, the general solution is in the form

$$y = e^{\alpha x}(C_1 \cos \beta x + C_2 \sin \beta x)$$

Then,

$$\alpha = -a/2 = -1/2$$

$$\beta = \sqrt{\frac{4b - a^2}{2}} = \sqrt{\frac{(4)(1) - (1)^2}{2}} = \frac{\sqrt{3}}{2}$$

Therefore, the general solution is

$$y = e^{-\frac{1}{2}x}\left(C_1 \cos \frac{\sqrt{3}}{2}x + C_2 \sin \frac{\sqrt{3}}{2}x\right)$$

The answer is (A).

7. Multiplying the equation by 2 gives

$$x'' + 8x' + 16x = 10$$

The characteristic equation is

$$r^2 + 8r + 16 = 0$$

The roots of the characteristic equation are

$$r_1 = r_2 = -4$$

The homogeneous (natural) response is

$$x_{\text{natural}} = Ae^{-4t} + Bte^{-4t}$$

By inspection, $x = 5/8$ is a particular solution that solves the nonhomogeneous equation, so the total response is

$$x = Ae^{-4t} + Bte^{-4t} + \tfrac{5}{8}$$

Since $x = 1$ at $t = 0$,

$$1 = Ae^0 + \tfrac{5}{8}$$
$$A = \tfrac{3}{8}$$

Differentiating x,

$$x' = \tfrac{3}{8}(-4)e^{-4t} + B(-4te^{-4t} + e^{-4t}) + 0$$

Since $x' = 0$ at $t = 0$,

$$0 = -\tfrac{3}{2} + B(0 + 1)$$
$$B = \tfrac{3}{2}$$
$$x = \tfrac{3}{8}e^{-4t} + \tfrac{3}{2}te^{-4t} + \tfrac{5}{8}$$

The answer is (D).

8. This is a first-order, linear, homogeneous differential equation with characteristic equation $r + 4 = 0$.

$$x' + 4x = 0$$
$$x = x_0 e^{-4t}$$
$$x(0) = x_0 e^{(-4)(0)}$$
$$= 12$$
$$x_0 = 12$$
$$x = 12e^{-4t}$$
$$x(2) = 12e^{(-4)(2)}$$
$$= 12e^{-8}$$
$$= 4.03 \times 10^{-3} \quad (4.0 \times 10^{-3})$$

The answer is (B).

9. By inspection, $f(t) = t$, with the period $T = 1$. The angular frequency is

$$\omega_0 = \frac{2\pi}{T} = \frac{2\pi}{1} = 2\pi$$

The average is

$$a_0 = (1/T) \int_0^T f(t)\, dt = (1/T) \int_0^T t\, dt = \frac{1}{2} t^2 \Big|_0^1 = \frac{1}{2} - 0$$
$$= \frac{1}{2}$$

The general a term is

$$a_n = (2/T) \int_0^T f(t) \cos(n\omega_0 t)\, dt$$
$$= 2 \int_0^1 t \cos(2\pi nt)\, dt$$
$$= 0$$

The general b term is

$$b_n = (2/T) \int_0^T f(t) \sin(n\omega_0 t)\, dt$$
$$= 2 \int_0^1 t \sin(2\pi nt)\, dt$$
$$= \frac{-1}{\pi n}$$

The answer is (B).

10. The value of a number in a Fibonacci sequence is the sum of the previous two numbers in the sequence.

Use the second-order difference equation.

$$f(k) = f(k-1) + f(k-2)$$
$$f(3) = f(2) + f(1) = 1.3 + 4$$
$$= 5.3$$
$$f(4) = f(3) + f(2) = 5.3 + 1.3$$
$$= 6.6$$

The answer is (D).

Probability and Statistics

PRACTICE PROBLEMS

1. What is the approximate probability that no people in a group of seven have the same birthday?

(A) 0.056

(B) 0.43

(C) 0.92

(D) 0.94

2. A study gives the following results for a total sample size of 12.

$$3, 4, 4, 5, 8, 8, 8, 10, 11, 15, 18, 20$$

What is most nearly the mean?

(A) 8.9

(B) 9.5

(C) 11

(D) 12

3. A study gives the following results for a total sample size of 8.

$$2, 3, 5, 8, 8, 10, 10, 12$$

The mean of the sample is 7.25. What is most nearly the standard deviation?

(A) 2.5

(B) 2.9

(C) 3.3

(D) 3.7

4. A study gives the following results for a total sample size of 6.

$$10, 12, 13, 14, 14, 15$$

The mean of the sample is 13. What is most nearly the sample standard deviation?

(A) 0.85

(B) 0.90

(C) 1.6

(D) 1.8

5. A study has a sample size of 5, a standard deviation of 10.4, and a sample standard deviation of 11.6. What is most nearly the variance?

(A) 46

(B) 52

(C) 110

(D) 130

6. A study has a sample size of 9, a standard deviation of 4.0, and a sample standard deviation of 4.2. What is most nearly the sample variance?

(A) 16

(B) 18

(C) 34

(D) 36

7. A bag contains 100 balls numbered 1 to 100. One ball is drawn from the bag. What is the probability that the number on the ball selected will be odd or greater than 80?

(A) 0.1

(B) 0.5

(C) 0.6

(D) 0.7

8. Measurements of the water content of soil from a borrow site are normally distributed with a mean of 14.2% and a standard deviation of 2.3%. What is the probability that a sample taken from the site will have a water content above 16% or below 12%?

(A) 0.13

(B) 0.25

(C) 0.37

(D) 0.42

9. What is the probability that either exactly two heads or exactly three heads will be thrown if six fair coins are tossed at once?

(A) 0.35

(B) 0.55

(C) 0.59

(D) 0.63

10. Which of the following properties of probability is NOT valid?

(A) The probability of an event is always positive and less than or equal to one.

(B) The probability of an event which cannot occur in the population being examined is zero.

(C) If events A and B are mutually exclusive, then the probability of either event occurring in the same population is zero.

(D) The probability of two events, A and B, both occurring is $P(A+B) = P(A) + P(B) - P(A, B)$.

11. One fair die is used in a dice game. A player wins \$10 if he rolls either a 1 or a 6. He loses \$5 if he rolls any other number. What is the expected winning for one roll of the die?

(A) \$0.00

(B) \$3.30

(C) \$5.00

(D) \$6.70

12. A simulation model for a transportation system is run for 30 replications, and the mean percentage utilization of the transporter used by the system is recorded for each replication. Those 30 data points are then used to form a confidence interval on mean transporter utilization for the system. At a 95% confidence level, the confidence interval is found to be 37.2% ± 3.4%.

Given this information, which of the following facts can be definitively stated about the system?

(A) At 95% confidence, the sample mean of transporter utilization lies in the range 37.2% ± 3.4%.

(B) At 95% confidence, the population mean of transporter utilization lies in the range 37.2% ± 3.4%.

(C) At 95% confidence, the population mean of transporter utilization lies outside of the range of 37.2% ± 3.4%.

(D) At 5% confidence, the population mean of transporter utilization lies inside of the range of 37.2% ± 3.4%.

13. What is the approximate probability of exactly two people in a group of seven having a birthday on April 15?

(A) 1.2×10^{-18}

(B) 2.4×10^{-17}

(C) 7.4×10^{-6}

(D) 1.6×10^{-4}

14. What are the arithmetic mean and sample standard deviation of the following numbers?

$$71.3, 74.0, 74.25, 78.54, 80.6$$

(A) 74.3, 2.7

(B) 74.3, 3.7

(C) 75.0, 2.7

(D) 75.7, 3.8

15. Four fair coins are tossed at once. What is the probability of obtaining three heads and one tail?

(A) 1/4 (0.25)

(B) 3/8 (0.375)

(C) 1/2 (0.50)

(D) 3/4 (0.75)

16. Set A and set B are subsets of set Q. The values within each set are shown.

$$A = (4, 7, 9)$$
$$B = (4, 5, 9, 10)$$
$$Q = (4, 5, 6, 7, 8, 9, 10)$$

What is the union of the complement of set A and set B, $\overline{A} \cup B$?

(A) $(4, 5, 6, 7, 8, 9, 10)$

(B) $(4, 5, 7, 9, 10)$

(C) $(4, 5, 6, 8, 9, 10)$

(D) $(5, 10)$

17. Set A consists of elements $(1, 3, 6)$, and set B consists of elements $(1, 2, 6, 7)$. Both sets come from the universe of $(1, 2, 3, 4, 5, 6, 7, 8)$. What is the intersection, $\overline{A} \cap B$?

(A) $(2, 7)$

(B) $(2, 3, 7)$

(C) $(2, 4, 5, 7, 8)$

(D) $(4, 5, 8)$

SOLUTIONS

1. This is the classic "birthday problem." The problem is to find the probability that all seven people have distinctly different birthdays. The solution can be found from simple counting.

The first person considered can be born on any day, which means the probability they will not be born on one of the 365 days of the year is 0.

$$P(1) = 1 - P(\text{not } 1) = 1 - 0 = 1 \quad (365/365)$$

The probability the second person will be born on the same day as the first person is 1 in 365. (The second person can be born on any other of the 364 days.) The probability that the second person is born on any other day is

$$P(2) = 1 - P(\text{not } 2) = 1 - \frac{1}{365} = \frac{364}{365}$$

The third person cannot have been born on either of the same days as the first and second people, which has a 2 in 365 probability of happening. The probability that the third person is born on any other day is

$$P(3) = 1 - P(\text{not } 3) = 1 - \frac{2}{365} = \frac{363}{365}$$

This logic continues to the seventh person. The probability that all seven conditions are simultaneously satisfied is

$$P(7 \text{ distinct birthdays})$$
$$= P(1) \times P(2) \times P(3) \times P(4) \times P(5)$$
$$\times P(6) \times P(7)$$
$$= \left(\frac{365}{365}\right)\left(\frac{364}{365}\right)\left(\frac{363}{365}\right)\left(\frac{362}{365}\right)\left(\frac{361}{365}\right)$$
$$\times \left(\frac{360}{365}\right)\left(\frac{359}{365}\right)$$
$$= 0.9438 \quad (0.94)$$

The answer is (D).

2. The mean is

$$\overline{X} = (1/n)\sum_{i=1}^{n} X_i$$
$$= \left(\frac{1}{12}\right)\left(\begin{array}{c} 3+4+4+5 \\ +8+8+8+10 \\ +11+15+18+20 \end{array}\right)$$
$$= 9.5$$

The answer is (B).

3. The standard deviation is calculated using the sample mean as an unbiased estimator of the population mean.

$$\sigma = \sqrt{(1/N)\sum(X_i - \mu)^2} \approx \sqrt{(1/N)\sum(X_i - \overline{X})^2}$$
$$= \sqrt{\left(\frac{1}{8}\right)\left(\begin{array}{c} (2-7.25)^2 + (3-7.25)^2 \\ +(5-7.25)^2 + (8-7.25)^2 \\ +(8-7.25)^2 + (10-7.25)^2 \\ +(10-7.25)^2 + (12-7.25)^2 \end{array}\right)}$$
$$= 3.34 \quad (3.3)$$

The answer is (C).

4. The sample standard deviation is

$$s = \sqrt{[1/(n-1)]\sum_{i=1}^{n}(X_i - \overline{X})^2}$$
$$= \sqrt{\left(\frac{1}{6-1}\right)\left(\begin{array}{c} (10-13)^2 + (12-13)^2 \\ +(13-13)^2 + (14-13)^2 \\ +(14-13)^2 + (15-13)^2 \end{array}\right)}$$
$$= 1.79 \quad (1.8)$$

The answer is (D).

5. The variance is

$$\sigma^2 = (10.4)^2 = 108 \quad (110)$$

The answer is (C).

6. The sample variance is

$$s^2 = (4.2)^2 = 17.64 \quad (18)$$

The answer is (B).

7. There are 50 odd-numbered balls. Including ball 100, there are 20 balls with numbers greater than 80.

$$P(A) = P(\text{ball is odd}) = \frac{50}{100} = 0.5$$
$$P(B) = P(\text{ball} > 80) = \frac{20}{100} = 0.2$$

It is possible for the number on the selected ball to be both odd and greater than 80. Use the law of total probability.

$$P(A + B) = P(A) + P(B) - P(A, B)$$
$$= P(A) + P(B) - P(A)P(B)$$
$$P(\text{odd or} > 80) = 0.5 + 0.2 - (0.5)(0.2) = 0.6$$

The answer is (C).

8. Find the standard normal values for the two points of interest.

$$Z_{16\%} = \frac{x - \mu}{\sigma} = \frac{16\% - 14.2\%}{2.3\%}$$
$$= 0.78 \quad [\text{use } 0.80]$$

$$Z_{12\%} = \frac{x - \mu}{\sigma} = \frac{12\% - 14.2\%}{2.3\%}$$
$$= -0.96 \quad [\text{use } -1.00]$$

Use the unit normal distribution table. The probabilities being sought can be found from the values of $R(x)$ for both standard normal values. $R(0.80) = 0.2119$ and $R(1.00) = 0.1587$. The probability that the sample will fall outside these values is the sum of the two values.

$$P(x < 12\% \text{ or } x > 16\%) = 0.2119 + 0.1587$$
$$= 0.3706 \quad (0.37)$$

The answer is (C).

9. Find the probability of exactly 2 heads being thrown. The probability will be the quotient of the total number of possible combinations of six objects taken two at a time and the total number of possible outcomes from tossing six fair coins. The total number of possible outcomes is $(2)^6 = 64$. The total number of possible combinations in which exactly two heads are thrown is

$$C(n, r) = \frac{n!}{r!(n - r)!} = \frac{6!}{2!(6 - 2)!}$$
$$= 15$$

The probability of exactly two heads out of six fair coins is

$$P(A) = P(2 \text{ heads}) = \frac{15}{64} = 0.234$$

The probability of exactly three heads being thrown is found similarly. The total number of possible combinations in which exactly three heads are thrown is

$$C(n, r) = \frac{n!}{r!(n - r)!} = \frac{6!}{3!(6 - 3)!}$$
$$= 20$$

The probability of exactly three heads out of six fair coins is

$$P(B) = P(3 \text{ heads}) = \frac{20}{64} = 0.313$$

From the law of total probability, the probability that either of these outcomes will occur is the sum of the individual probabilities that the outcomes will occur, minus the probability that both will occur. These two

outcomes are mutually exclusive (i.e., both cannot occur), so the probability of both happening is 0.

The total probability is

$$P(2 \text{ heads or } 3 \text{ heads}) = P(A) + P(B) - P(A, B)$$
$$= 0.234 + 0.313 - 0$$
$$= 0.547 \quad (0.55)$$

The answer is (B).

10. If events A and B are mutually exclusive, the probability of both occurring is zero. However, either event could occur by itself, and the probability of that is nonzero.

The answer is (C).

11. For a fair die, the probability of any face turning up is $^1/_6$. There are two ways to win, and there are four ways to lose. The expected value is

$$E[X] = \sum_{k=1}^{n} x_k f(x_k) = (\$10)\left((2)\left(\frac{1}{6}\right)\right) + (-\$5)\left((4)\left(\frac{1}{6}\right)\right)$$
$$= \$0.00$$

The answer is (A).

12. A 95% confidence interval on mean transporter utilization means there is a 95% chance the population (or true) mean transporter utilization lies within the given interval.

The answer is (B).

13. Use the binomial probability function to calculate the probability that two of the seven samples will have been born on April 15. $x = 2$, and the sample size, n, is 7.

The probability that a person will have been born on April 15 is 1/365. Therefore, the probability of "success," p, is 1/365, and the probability of "failure," $q = 1 - p$, is 364/365.

$$P_n(x) = \frac{n!}{x!(n - x)!} p^x q^{n-x}$$
$$P_7(2) = \left(\frac{7!}{2!(7 - 2)!}\right)\left(\frac{1}{365}\right)^2 \left(\frac{364}{365}\right)^{7-2}$$
$$= (21)\left(\frac{1}{365}\right)^2 \left(\frac{364}{365}\right)^5$$
$$= 1.555 \times 10^{-4} \quad (1.6 \times 10^{-4})$$

The answer is (D).

14. The arithmetic mean is

$$\overline{X} = (1/n)\sum_{i=1}^{n} X_i$$

$$= \left(\frac{1}{5}\right)(71.3 + 74.0 + 74.25 + 78.54 + 80.6)$$

$$= 75.738 \quad (75.7)$$

The sample standard deviation is

$$s = \sqrt{[1/(n-1)]\sum_{i=1}^{n}(X_i - \overline{X})^2}$$

$$= \sqrt{\left(\frac{1}{5-1}\right)\left(\begin{array}{c}(71.3 - 75.738)^2 + (74.0 - 75.738)^2 \\ + (74.25 - 75.738)^2 \\ + (78.54 - 75.738)^2 \\ + (80.6 - 75.738)^2\end{array}\right)}$$

$$= 3.756 \quad (3.8)$$

The answer is (D).

15. The binomial probability function can be used to determine the probability of three heads in four trials.

$$p = P(\text{heads}) = 0.5$$

$$q = P(\text{not heads}) = 1 - 0.5 = 0.5$$

$$n = \text{number of trials} = 4$$

$$x = \text{number of successes} = 3$$

From the binomial function,

$$P_n(x) = \frac{n!}{x!(n-x)!}p^x q^{n-x}$$

$$= \left(\frac{4!}{3!(4-3)!}\right)(0.5)^3(0.5)^{4-3}$$

$$= 0.25 \quad (1/4)$$

The answer is (A).

16. The complement of set A contains all of the members of set Q that are not members of set A: $(5, 6, 8, 10)$.

The union of the complement of set A and set B is the set of all members appearing in either.

$$\overline{A} \cup B = (5, 6, 8, 10) \cup (4, 5, 9, 10)$$

$$= (4, 5, 6, 8, 9, 10)$$

The answer is (C).

17. Set "not A" consists of all universe elements not in set A: $(2, 4, 5, 7, 8)$.

The intersection of "not A" and B is the set of all elements appearing in both.

$$\overline{A} \cap B = (2, 4, 5, 7, 8) \cap (1, 2, 6, 7)$$

$$= (2, 7)$$

The answer is (A).

6 Fluid Properties

PRACTICE PROBLEMS

1. A leak from a faucet comes out in separate drops instead of a stream. The main cause of this phenomenon is

(A) gravity

(B) air resistance

(C) viscosity

(D) surface tension

2. A solid cylinder is concentric with a straight pipe. The cylinder is 0.5 m long and has an outside diameter of 8 cm. The pipe has an inside diameter of 8.5 cm. The annulus between the cylinder and the pipe contains stationary oil. The oil has a specific gravity of 0.92 and a kinematic viscosity of 5.57×10^{-4} m^2/s. The force needed to move the cylinder along the pipe at a constant velocity of 1 m/s is most nearly

(A) 5.9 N

(B) 12 N

(C) 26 N

(D) 55 N

3. Kinematic viscosity can be expressed in

(A) m^2/s

(B) s^2/m

(C) kg·s^2/m

(D) kg/s

4. Which three of the following must be satisfied by the flow of any fluid, whether real or ideal?

I.   Newton's second law of motion

II.   the continuity equation

III.   uniform velocity distribution

IV.   Newton's law of viscosity

V.   conservation of energy

(A) I, II, and III

(B) I, II, and V

(C) I, III, and V

(D) II, IV, and V

5. 15 kg of a fluid with a density of 790 kg/m^3 is mixed with 10 kg of water. The volumes are additive, and the resulting mixture is homogeneous. The specific volume of the resulting mixture is most nearly

(A) 0.0012 m^3/kg

(B) 0.0027 m^3/kg

(C) 0.0047 m^3/kg

(D) 0.0061 m^3/kg

6. The rise or fall of liquid in a small-diameter capillary tube is NOT affected by

(A) adhesive forces

(B) cohesive forces

(C) surface tension

(D) fluid viscosity

7. A capillary tube 3.8 mm in diameter is placed in a beaker of 40°C distilled water. The surface tension is 0.0696 N/m, and the angle made by the water with the wetted tube wall is negligible. The specific weight of water at this temperature is 9.730 kN/m^3. The height to which the water will rise in the tube is most nearly

(A) 1.2 mm

(B) 3.6 mm

(C) 7.5 mm

(D) 9.2 mm

SOLUTIONS

1. Surface tension is caused by the molecular cohesive forces in a fluid. It is the main cause of the formation of drops of water.

The answer is (D).

2. Treat the cylinder as a moving plate, and use Newton's law of viscosity. Find the absolute viscosity of the oil.

$$\nu = \frac{\mu}{\rho}$$

$$\mu = \nu\rho = \left(5.57 \times 10^{-4} \ \frac{m^2}{s}\right)(0.92)\left(1000 \ \frac{kg}{m^3}\right)$$

$$= 0.512 \ Pa{\cdot}s$$

The width of the separation between the cylinder and the pipe is

$$\delta = \frac{d_{pipe} - d_{cylinder}}{2} = \frac{8.5 \ cm - 8 \ cm}{2}$$

$$= 0.25 \ cm$$

The interval surface area of the cylinder is

$$A = \pi dL = \pi \left(\frac{8 \ cm}{100 \ \frac{cm}{m}}\right)(0.5 \ m)$$

$$= 0.126 \ m^2$$

Find the force needed.

$$\frac{F}{A} = \mu\left(\frac{d\mathrm{v}}{dy}\right)$$

$$F = A\mu\left(\frac{d\mathrm{v}}{dy}\right) \approx A\mu\left(\frac{\Delta\mathrm{v}}{\Delta y}\right)$$

$$= (0.126 \ m^2)(0.512 \ Pa{\cdot}s)\left(\frac{1 \ \frac{m}{s}}{0.25 \ cm}\right)\left(100 \ \frac{cm}{m}\right)$$

$$= 25.8 \ N \quad (26 \ N)$$

The answer is (C).

3. Typical units of kinematic viscosity are m^2/s.

The answer is (A).

4. Newton's second law, the continuity equation, and the principle of conservation of energy always apply for any fluid.

The answer is (B).

5. Calculate the volumes. Use a standard water density of 1000 kg/m^3.

$$\rho = \frac{m}{V}$$

$$V = \frac{m}{\rho}$$

$$V_{water} = \frac{m}{\rho} = \frac{10 \ kg}{1000 \ \frac{kg}{m^3}} = 0.010 \ m^3$$

$$V_{fluid} = \frac{m}{\rho} = \frac{15 \ kg}{790 \ \frac{kg}{m^3}} = 0.019 \ m^3$$

The total volume is

$$V_{total} = V_{water} + V_{fluid} = 0.010 \ m^3 + 0.019 \ m^3$$

$$= 0.029 \ m^3$$

The density of the mixture is the total mass divided by the total volume.

$$\rho_{mixture} = \frac{m_{water} + m_{fluid}}{V_{total}} = \frac{10 \ kg + 15 \ kg}{0.029 \ m^3}$$

$$= 862 \ kg/m^3$$

The specific volume of the mixture is the reciprocal of its density.

$$v_{mixture} = \frac{1}{\rho_{mixture}} = \frac{1}{862 \ \frac{kg}{m^3}}$$

$$= 0.00116 \ m^3/kg \quad (0.0012 \ m^3/kg)$$

The answer is (A).

6. The height of capillary rise is

$$h = 4\sigma \cos\beta/\gamma d$$

σ is the surface tension of the fluid, β is the angle of contact, γ is the specific weight of the liquid, and d is the diameter of the tube.

The viscosity of the fluid is not directly relevant to the height of capillary rise.

The answer is (D).

7. Since the contact angle is neglible, use 0° for β. The capillary rise in liquid is

$$h = 4\sigma \cos\beta/\gamma d$$

$$= \frac{(4)\left(0.0696 \ \frac{N}{m}\right)\cos 0°\left(1000 \ \frac{mm}{m}\right)}{\left(9.730 \ \frac{kN}{m^3}\right)\left(1000 \ \frac{N}{kN}\right)(3.8 \ mm)}$$

$$= 7.53 \times 10^{-3} \ m \quad (7.5 \ mm)$$

The answer is (C).

7 Fluid Statics

PRACTICE PROBLEMS

1. A barometer contains mercury with a density of $13\,600$ kg/m^3. Atmospheric conditions are 95.8 kPa and 20°C. At 20°C, the vapor pressure of the mercury is 0.000173 kPa. The column of mercury will rise to a height of most nearly

(A) 0.38 m

(B) 0.48 m

(C) 0.72 m

(D) 0.82 m

2. The manometer shown contains water, mercury, and glycerine. The specific gravity of mercury is 13.6, and the specific gravity of glycerine is 1.26.

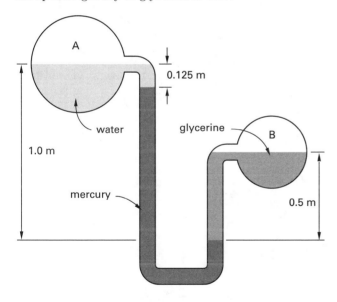

What is most nearly the difference in pressure between points A and B?

(A) 35 kPa

(B) 42 kPa

(C) 55 kPa

(D) 110 kPa

3. An open water manometer is used to measure the pressure in a tank. The tank is cylindrical with hemispherical ends. The tank is half-filled with 50 000 kg of a liquid chemical that is not miscible in water. The manometer tube is filled with liquid chemical up to the water.

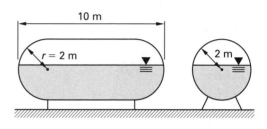

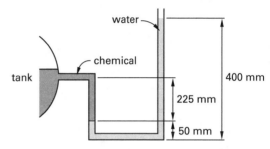

What is most nearly the pressure in the tank relative to the atmospheric pressure?

(A) 1.4 kPa

(B) 1.9 kPa

(C) 2.4 kPa

(D) 3.4 kPa

4. A pressure vessel is connected to a simple U-tube open to the atmosphere as shown. A 10 cm deflection of mercury is observed. The density of mercury is $13\,600 \text{ kg/m}^3$. The atmospheric pressure is 101 kPa.

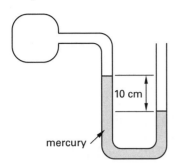

mercury

The vacuum within the vessel is most nearly

(A) 1.0 kPa

(B) 13 kPa

(C) 39 kPa

(D) 78 kPa

5. A tank contains a gate 2 m tall and 5 m wide as shown. The tank is filled with water to a depth of 10 m.

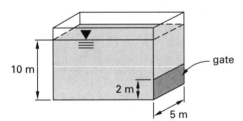

gate

10 m

2 m

5 m

The total force on the gate is most nearly

(A) 90 kN

(B) 440 kN

(C) 880 kN

(D) 980 kN

6. A 1 m × 2 m inclined plate is submerged as shown.

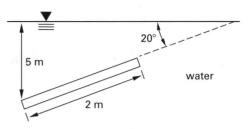

20°

5 m

water

2 m

The normal force acting on the upper surface of the plate is most nearly

(A) 32 kN

(B) 56 kN

(C) 68 kN

(D) 91 kN

7. The water tank shown has a width of 0.3 m. The rounded corner has a radius of 0.9 m.

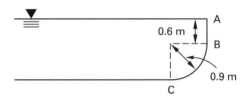

A

0.6 m

B

0.9 m

C

The magnitude and direction (in degrees from the horizontal) of the water force over the length of wall from point A to point C is most nearly

(A) 3400 N, 46°

(B) 3400 N, 73°

(C) 4800 N, 46°

(D) 4800 N, 73°

SOLUTIONS

1. Find the height of the mercury in the column.

$$p_A = p_B + \rho g h$$

$$h = \frac{p_A - p_B}{\rho g}$$

$$= \frac{(95.8 \text{ kPa} - 0.000173 \text{ kPa})\left(1000 \frac{\text{Pa}}{\text{kPa}}\right)}{\left(13\,600 \frac{\text{kg}}{\text{m}^3}\right)\left(9.81 \frac{\text{m}}{\text{s}^2}\right)}$$

$$= 0.7181 \text{ m} \quad (0.72 \text{ m})$$

The answer is (C).

2. The manometer can be labeled as shown.

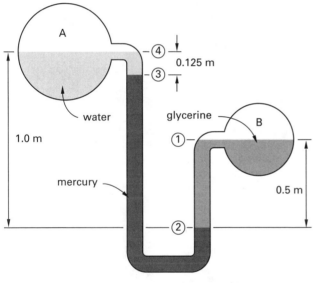

The pressure at level 2 is the same in both (left and right) legs of the manometer. For the left leg,

$$p_2 = p_A + \rho_{\text{water}} g h_{3\text{-}4} + \rho_{\text{mercury}} g h_{2\text{-}3}$$

For the right leg,

$$p_2 = p_B + \rho_{\text{glycerine}} g h_{1\text{-}2}$$

Equating these two equations for p_2 and solving for the pressure difference $p_A - p_B$ gives

$$p_A - p_B = g(\rho_{\text{glycerine}} h_{1\text{-}2} - \rho_{\text{water}} h_{3\text{-}4} - \rho_{\text{Hg}} h_{2\text{-}3})$$

$$= g\rho_{\text{water}}\left(\begin{array}{c} SG_{\text{glycerine}} h_{1\text{-}2} - SG_{\text{water}} h_{3\text{-}4} \\ - SG_{\text{Hg}} h_{2\text{-}3} \end{array}\right)$$

$$= \left(9.81 \frac{\text{m}}{\text{s}^2}\right)\left(1000 \frac{\text{kg}}{\text{m}^3}\right)$$

$$\times \left(\begin{array}{c} (1.26)(0.5 \text{ m}) - (1.00)(0.125 \text{ m}) \\ - (13.6)(1.0 \text{ m} - 0.125 \text{ m}) \end{array}\right)$$

$$= -111\,785 \text{ Pa} \quad (110 \text{ kPa})$$

The answer is (D).

3. Calculate the density of the chemical from the volume and mass. The total volume of the tank is

$$V = \tfrac{4}{3}\pi r^3 + \pi r^2 (L - 2r)$$

$$= \tfrac{4}{3}\pi (2 \text{ m})^3 + \pi (2 \text{ m})^2 \left(10 \text{ m} - (2)(2 \text{ m})\right)$$

$$= 108.9 \text{ m}^3$$

The contents have a mass of $50\,000$ kg, and the tank is half full, so the density of the chemical is

$$\rho_{\text{chemical}} = \frac{m}{V} = \frac{50\,000 \text{ kg}}{\left(\frac{1}{2}\right)(108.9 \text{ m}^3)}$$

$$= 918.2 \text{ kg/m}^3$$

The relative pressure is

$$p_0 - p_2 = \rho_{\text{water}} g h_2 - \rho_{\text{chemical}} g h_1$$

$$= \left(1000 \frac{\text{kg}}{\text{m}^3}\right)\left(9.81 \frac{\text{m}}{\text{s}^2}\right)\left(\frac{400 \text{ mm} - 50 \text{ mm}}{1000 \frac{\text{mm}}{\text{m}}}\right)$$

$$- \left(918.2 \frac{\text{kg}}{\text{m}^3}\right)\left(9.81 \frac{\text{m}}{\text{s}^2}\right)\left(\frac{225 \text{ m}}{1000 \frac{\text{mm}}{\text{m}}}\right)$$

$$= 1407 \text{ Pa} \quad (1.4 \text{ kPa})$$

The answer is (A).

Fluid Mechanics

4. The pressure is

$$p_{\text{gage}} = -\rho g h$$

$$= -\left(13\,600\ \frac{\text{kg}}{\text{m}^3}\right)\left(9.81\ \frac{\text{m}}{\text{s}^2}\right)\left(\frac{10\ \text{cm}}{100\ \frac{\text{cm}}{\text{m}}}\right)$$

$$= -13\,342\ \text{Pa} \quad (13\ \text{kPa})$$

The answer is (B).

5. $h_1 = 10\ \text{m} - 2\ \text{m} = 8\ \text{m}$. The average pressure is

$$\overline{p} = \tfrac{1}{2}\rho g(h_1 + h_2)$$

$$= \left(\tfrac{1}{2}\right)\left(1000\ \frac{\text{kg}}{\text{m}^3}\right)\left(9.81\ \frac{\text{m}}{\text{s}^2}\right)(8\ \text{m} + 10\ \text{m})$$

$$= 88\,290\ \text{Pa}$$

The total force acting on the gate is

$$R = \overline{p}A = (88\,290\ \text{Pa})\Big((2\ \text{m})(5\ \text{m})\Big)$$

$$= 882\,900\ \text{N} \quad (880\ \text{kN})$$

The answer is (C).

6. The upper edge of the plate is at a depth of

$$h_1 = 5\ \text{m} - (2\ \text{m})\sin 20° = 4.32\ \text{m}$$

The average pressure is

$$\overline{p} = \tfrac{1}{2}\rho g(h_1 + h_2)$$

$$= \left(\tfrac{1}{2}\right)\left(1000\ \frac{\text{kg}}{\text{m}^3}\right)\left(9.81\ \frac{\text{m}}{\text{s}^2}\right)(4.32\ \text{m} + 5\ \text{m})$$

$$= 45\,695\ \text{Pa}$$

The normal force acting on the plate is

$$R = \overline{p}A = (45\,695\ \text{Pa})\Big((1\ \text{m})(2\ \text{m})\Big)$$

$$= 91\,390\ \text{N} \quad (91\ \text{kN})$$

The answer is (D).

7. Find separately the horizontal and vertical components of the force acting on the wall from point A to point C. For the horizontal component, $h_1 = 0$ m, and $h_2 = 0.6$ m $+ 0.9$ m $= 1.5$ m.

$$\overline{p}_x = \tfrac{1}{2}\rho g(h_1 + h_2)$$

$$= \left(\tfrac{1}{2}\right)\left(1000\ \frac{\text{kg}}{\text{m}^3}\right)\left(9.81\ \frac{\text{m}}{\text{s}^2}\right)(0\ \text{m} + 1.5\ \text{m})$$

$$= 7358\ \text{Pa}$$

The horizontal component of the force is

$$R_x = \overline{p}_x A = (7358\ \text{Pa})\Big((1.5\ \text{m})(0.3\ \text{m})\Big)$$

$$= 3311\ \text{N}$$

For the vertical component, calculate the weight of the water above the section of wall from point B to point C. The volume consists of rectangular prism $0.3\,\text{m} \times 0.6\ \text{m} \times 0.9$ m plus a quarter of a cylinder, which has a radius of 0.9 m and a length of 0.3 m.

$$V = V_1 + V_2$$

$$= (0.3\ \text{m})(0.6\ \text{m})(0.9\ \text{m}) + \frac{\pi(0.9\ \text{m})^2(0.3\ \text{m})}{4}$$

$$= 0.3529\ \text{m}^3$$

The vertical component of the force equals the weight of this volume of water.

$$R_y = \rho g V$$

$$= \left(1000\ \frac{\text{kg}}{\text{m}^3}\right)\left(9.81\ \frac{\text{m}}{\text{s}^2}\right)(0.3529\ \text{m}^3)$$

$$= 3461\ \text{N}$$

The resultant force acting on this section of wall is

$$R = \sqrt{R_x^2 + R_y^2} = \sqrt{(3311\ \text{N})^2 + (3461\ \text{N})^2}$$

$$= 4790\ \text{N} \quad (4800\ \text{N})$$

The direction of the resultant force from the horizontal is

$$\theta = \arctan \frac{R_y}{R_x} = \arctan \frac{3461\ \text{N}}{3311\ \text{N}}$$

$$= 46.27° \quad (46°)$$

The answer is (C).

8 Fluid Measurement and Similitude

PRACTICE PROBLEMS

1. A pitot tube is used to measure the flow of an incompressible fluid with a density of 926 kg/m^3. The velocity is measured as 2 m/s, and the stagnation pressure is 14.1 kPa. The static pressure of the fluid where the measurement is taken is most nearly

(A) 10.4 kPa

(B) 11.7 kPa

(C) 12.2 kPa

(D) 13.5 kPa

2. A sharp-edged orifice with a 50 mm diameter opening in the vertical side of a large tank discharges under a head of 5 m. The coefficient of contraction is 0.62, and the coefficient of velocity is 0.98.

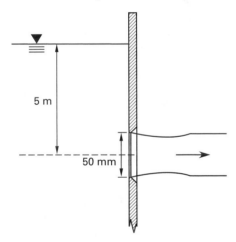

The rate of discharge is most nearly

(A) 0.00031 m^3/s

(B) 0.0040 m^3/s

(C) 0.010 m^3/s

(D) 0.012 m^3/s

3. The velocity of the water in the stream shown is 1.2 m/s.

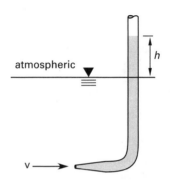

The height of water in the pitot tube is most nearly

(A) 3.7 cm

(B) 4.6 cm

(C) 7.3 cm

(D) 9.2 cm

4. A horizontal venturi meter with a diameter of 15 cm at the throat is installed in a 45 cm water main. A differential manometer gauge is partly filled with mercury (the remainder of the tube is filled with water) and connected to the meter at the throat and inlet. The mercury column stands 37.5 cm higher in one leg than in the other. The specific gravity of mercury is 13.6.

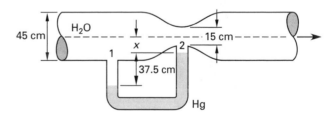

Neglecting friction, the flow through the meter is most nearly

(A) 0.10 m^3/s

(B) 0.17 m^3/s

(C) 0.23 m^3/s

(D) 0.28 m^3/s

5. A 1:1 model of a torpedo is tested in a wind tunnel according to the Reynolds number similarity. At the testing temperature, the kinematic viscosity of air is 1.41×10^{-5} m²/s, and the kinematic viscosity of water is 1.31×10^{-6} m²/s. If the velocity of the torpedo in water is 7 m/s, the air velocity in the wind tunnel should be most nearly

(A) 0.62 m/s

(B) 7.0 m/s

(C) 18 m/s

(D) 75 m/s

6. A 2 m tall, 0.5 m inside diameter tank is filled with water. A 10 cm hole is opened 0.75 m from the bottom of the tank. Ignoring all orifice losses, the velocity of the exiting water is most nearly

(A) 4.75 m/s

(B) 4.80 m/s

(C) 4.85 m/s

(D) 4.95 m/s

7. Water flows from one reservoir to another through a perfectly insulated pipe. Between the two reservoirs, 100 m of head is lost due to friction. Water has a specific heat of 4180 J/kg·K. The increase in water temperature between the reservoirs is most nearly

(A) 0.23°C

(B) 0.52°C

(C) 0.70°C

(D) 1.0°C

SOLUTIONS

1. Solve the equation for velocity in a pitot tube for the static pressure.

$$v = \sqrt{\left(\frac{2}{\rho}\right)(p_0 - p_s)}$$

$$p_s = p_0 - \frac{\rho v^2}{2}$$

$$= 14.1 \text{ kPa} - \frac{\left(926 \frac{\text{kg}}{\text{m}^3}\right)\left(2 \frac{\text{m}}{\text{s}}\right)^2}{(2)\left(1000 \frac{\text{Pa}}{\text{kPa}}\right)}$$

$$= 12.2 \text{ kPa}$$

The answer is (C).

2. The area of the opening is

$$A = \frac{\pi D^2}{4} = \frac{\pi (50 \text{ mm})^2}{(4)\left(1000 \frac{\text{mm}}{\text{m}}\right)^2}$$

$$= 0.00196 \text{ m}^2$$

The coefficient of discharge is

$$C = C_c C_v = (0.62)(0.98)$$

$$= 0.6076$$

The discharge rate is

$$Q = CA_0\sqrt{2gh}$$

$$= (0.6076)(0.00196 \text{ m}^2)\sqrt{(2)\left(9.81 \frac{\text{m}}{\text{s}^2}\right)(5 \text{ m})}$$

$$= 0.012 \text{ m}^3/\text{s}$$

The answer is (D).

3. The difference in height between the pitot tube and the free-water surface is a measure of the difference in static and stagnation pressures. Solve for the height of the water.

$$v = \sqrt{\left(\frac{2}{\rho}\right)(p_0 - p_s)} = \sqrt{\left(\frac{2}{\rho}\right)\rho gh}$$

$$= \sqrt{2gh}$$

$$h = \frac{v^2}{2g} = \frac{\left(1.2 \frac{\text{m}}{\text{s}}\right)^2}{(2)\left(9.81 \frac{\text{m}}{\text{s}^2}\right)}$$

$$= 0.073 \text{ m} \quad (7.3 \text{ cm})$$

The answer is (C).

4. The areas of the pipes are

$$A_1 = \frac{\pi D^2}{4} = \frac{\pi(45\text{ cm})^2}{(4)\left(100\ \frac{\text{cm}}{\text{m}}\right)^2} = 0.159\text{ m}^2$$

$$A_2 = \frac{\pi D^2}{4} = \frac{\pi(15\text{ cm})^2}{(4)\left(100\ \frac{\text{cm}}{\text{m}}\right)^2} = 0.0177\text{ m}^2$$

The equation for flow through a venturi meter can be written in terms of a manometer fluid reading. For horizontal flow, $z_1 = z_2$.

$$Q = \frac{C_{\text{v}} A_2}{\sqrt{1 - \left(\frac{A_2}{A_1}\right)^2}} \sqrt{2g\left(\frac{p_1}{\gamma} + z_1 - \frac{p_2}{\gamma} - z_2\right)}$$

$$= \left(\frac{C_{\text{v}} A_2}{\sqrt{1 - \left(\frac{A_2}{A_1}\right)^2}}\right) \sqrt{\frac{2g(\rho_m - \rho)h}{\rho}}$$

Because friction is to be neglected, $C_{\text{v}} = 1$. (For venturi meters, C_{v} is usually very close to one because the diameter changes are gradual and there is little friction loss.)

$$Q = \left(\frac{C_{\text{v}} A_2}{\sqrt{1 - \left(\frac{A_2}{A_1}\right)^2}}\right) \sqrt{\frac{2g(\rho_m - \rho)h}{\rho}}$$

$$= \left(\frac{(1)(0.0177\text{ m}^2)}{\sqrt{1 - \left(\frac{0.0177\text{ m}^2}{0.159\text{ m}^2}\right)^2}}\right)$$

$$\times \sqrt{\frac{(2)\left(9.81\ \frac{\text{m}}{\text{s}^2}\right)\left(1000\ \frac{\text{kg}}{\text{m}^3}\right)}{\left(1000\ \frac{\text{kg}}{\text{m}^3}\right)\left(100\ \frac{\text{cm}}{\text{m}}\right)}}$$

$$= 0.171\text{ m}^3/\text{s}\quad(0.17\text{ m}^3/\text{s})$$

The answer is (B).

5. From the Reynolds number similarity,

$$\left[\frac{F_I}{F_V}\right]_p = \left[\frac{F_I}{F_V}\right]_m = \left[\frac{\text{v}l\rho}{\mu}\right]_p = \left[\frac{\text{v}l\rho}{\mu}\right]_m = [\text{Re}]_p = [\text{Re}]_m$$

The scale is 1:1, so the lengths of the prototype and model are the same ($l_m = l_p$).

The similarity equation reduces to

$$\left(\frac{\text{v}\rho}{\mu}\right)_p = \left(\frac{\text{v}\rho}{\mu}\right)_m$$

$$\left(\frac{\text{v}}{\nu}\right)_p = \left(\frac{\text{v}}{\nu}\right)_m$$

$$\text{v}_m = \text{v}_p\left(\frac{\nu_m}{\nu_p}\right) = \left(7\ \frac{\text{m}}{\text{s}}\right)\left(\frac{1.41 \times 10^{-5}\ \frac{\text{m}^2}{\text{s}}}{1.31 \times 10^{-6}\ \frac{\text{m}^2}{\text{s}}}\right)$$

$$= 75.3\text{ m/s}\quad(75\text{ m/s})$$

The answer is (D).

6. The hydraulic head at the hole is

$$h = 2\text{ m} - 0.75\text{ m} = 1.25\text{ m}$$

For an orifice discharging freely into the atmosphere,

$$Q = CA_0\sqrt{2gh}$$

As orifice losses are neglected, $C = 1$. Dividing both sides by A_0 gives

$$\text{v} = C\sqrt{2gh} = 1\sqrt{(2)\left(9.81\ \frac{\text{m}}{\text{s}^2}\right)(1.25\text{ m})}$$

$$= 4.95\text{ m/s}$$

The answer is (D).

7. Convert the frictional head loss to specific energy loss.

$$\Delta E = h_f g = (100\text{ m})\left(9.81\ \frac{\text{m}}{\text{s}^2}\right)$$

$$= 981\text{ m}^2/\text{s}^2\quad(981\text{ J/kg})$$

The temperature increase is

$$\Delta T = \frac{\Delta E}{c_p} = \frac{981\ \frac{\text{J}}{\text{kg}}}{4180\ \frac{\text{J}}{\text{kg·K}}} = 0.2347\text{K}\quad(0.23°\text{C})$$

The answer is (A).

Fluid Mechanics

Hydrology

PRACTICE PROBLEMS

1. Two adjacent fields contribute runoff to a collector. Field 1 is 2 ac in size and has a runoff coefficient of 0.35. Field 2 is 4 ac in size and has a runoff coefficient of 0.65. The rainfall intensity of the storm after the time to concentration is 3.9 in/hr. The peak runoff is most nearly

(A) 8.7 ft^3/sec

(B) 10 ft^3/sec

(C) 13 ft^3/sec

(D) 16 ft^3/sec

2. The rational formula runoff coefficient of a 950 ft × 600 ft property is 0.35. A storm occurs with a rainfall intensity of 4.5 in/hr. The peak runoff from this property is most nearly

(A) 21 ft^3/sec

(B) 30 ft^3/sec

(C) 62 ft^3/sec

(D) 90 ft^3/sec

3. A watershed occupies a 70 ac site. 45 ac of the site have been cleared and are used for pasture land with a runoff coefficient of 0.13; 3 ac are occupied by farm buildings, a house, and paved surfaces and have a runoff coefficient of 0.75; the remaining 22 ac are woodland with a runoff coefficient of 0.20. The total time to concentration for the watershed is 30 min. The 20 yr storm is characterized by the intensity duration curve shown.

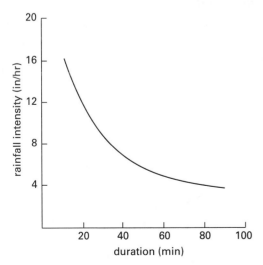

The peak runoff for the 20 yr storm is most nearly

(A) 50 ft^3/sec

(B) 110 ft^3/sec

(C) 240 ft^3/sec

(D) 530 ft^3/sec

4. The table shown contains curve numbers based on land use and soil type. A watershed contains 10 ac of residential land of soil type B and 5 ac of grassland of soil type A.

	soil type	
land use	type A	type B
residential	57	72
grassland	30	58

If the total precipitation is 11 in, what will be the approximate runoff from the watershed?

(A) 5.4 in

(B) 6.0 in

(C) 6.7 in

(D) 7.2 in

5. A drainage basin covers an area of 2.4 ac. During a storm with a sustained rainfall intensity of 0.6 in/hr, the peak runoff from the basin is 320 gal/min. What is most nearly the runoff coefficient for the basin?

(A) 0.38

(B) 0.50

(C) 0.65

(D) 0.85

SOLUTIONS

1. The runoff coefficients are given for each area. Combine the coefficients, weighting them by their respective contributing areas.

$$C = \frac{(2 \text{ ac})(0.35) + (4 \text{ ac})(0.65)}{2 \text{ ac} + 4 \text{ ac}}$$
$$= 0.55$$

Find the peak flow from the rational formula.

$$Q = CIA = (0.55)\left(3.9 \, \frac{\text{in}}{\text{hr}}\right)(6 \text{ ac})$$
$$= 12.87 \text{ ft}^3/\text{sec} \quad (13 \text{ ft}^3/\text{sec})$$

The answer is (C).

2. Use the rational formula to calculate the peak runoff from this property. To use this formula, the area of the property must be in acres.

$$A = \frac{(950 \text{ ft})(600 \text{ ft})}{43{,}560 \, \dfrac{\text{ft}^2}{\text{ac}}} = 13.09 \text{ ac}$$

The peak runoff is

$$Q = CIA = (0.35)\left(4.5 \, \frac{\text{in}}{\text{hr}}\right)(13.09 \text{ ac})$$
$$= 20.61 \text{ ft}^3/\text{sec} \quad (21 \text{ ft}^3/\text{sec})$$

The answer is (A).

3. The weighted average runoff coefficient for the watershed is

$$C = \frac{(0.13)(45 \text{ ac}) + (0.75)(3 \text{ ac}) + (0.20)(22 \text{ ac})}{70 \text{ ac}}$$
$$= 0.179$$

From the graph, at $t_c = 30$ min, $I = 9$ in/hr. Use the rational formula to calculate the peak runoff.

$$Q = CIA = (0.179)\left(9 \, \frac{\text{in}}{\text{hr}}\right)(70 \text{ ac})$$
$$= 112.5 \text{ ft}^3/\text{sec} \quad (110 \text{ ft}^3/\text{sec})$$

The answer is (B).

4. Find the weighted curve number for the watershed. From the table, the watershed contains 10 ac of land with a curve number of 72 and 5 ac of land with a curve number of 30.

$$CN = \frac{(10 \text{ ac})(72) + (5 \text{ ac})(30)}{10 \text{ ac} + 5 \text{ ac}} = 58$$

Find the storage capacity of the watershed.

$$S = \frac{1000}{CN} - 10 = \frac{1000}{58} - 10 = 7.241 \text{ in}$$

Find the runoff from 11 in of precipitation.

$$Q = \frac{(P - 0.2S)^2}{P + 0.8S} = \frac{(11 \text{ in} - (0.2)(7.241 \text{ in}))^2}{11 \text{ in} + (0.8)(7.241 \text{ in})}$$
$$= 5.433 \text{ in} \quad (5.4 \text{ in})$$

The answer is (A).

5. Convert the runoff to cubic feet per second.

$$Q = \frac{320 \, \dfrac{\text{gal}}{\text{min}}}{\left(7.48 \, \dfrac{\text{gal}}{\text{ft}^3}\right)\left(60 \, \dfrac{\text{sec}}{\text{min}}\right)} = 0.7130 \text{ ft}^3/\text{sec}$$

Use the rational formula to determine the runoff coefficient.

$$Q = CIA$$

$$C = \frac{Q}{IA} = \frac{0.7130 \, \dfrac{\text{ft}^3}{\text{sec}}}{\left(0.6 \, \dfrac{\text{in}}{\text{hr}}\right)(2.4 \text{ ac})}$$
$$= 0.4951 \quad (0.50)$$

The answer is (B).

10 Hydraulics

PRACTICE PROBLEMS

1. A model of a dam has been constructed so that the scale of dam to model is 15:1. The similarity is based on Froude numbers. At a certain point on the spillway of the model, the velocity is 5 m/s. At the corresponding point on the spillway of the actual dam, the velocity would most nearly be

(A) 6.7 m/s

(B) 7.5 m/s

(C) 15 m/s

(D) 19 m/s

2. An open channel has a cross-sectional area of flow of 0.5 m², a hydraulic radius of 0.15 m, and a roughness factor of 0.15. The slope of the hydraulic gradient needed to achieve a flow rate of 10 L/s is most nearly

(A) 1.1×10^{-4}

(B) 6.7×10^{-4}

(C) 1.1×10^{-3}

(D) 6.7×10^{-3}

3. Water flows through a converging channel as shown and discharges freely to the atmosphere at the exit. Flow is incompressible, and friction is negligible.

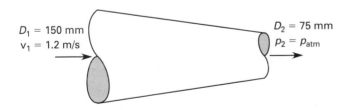

The gage pressure at the inlet is most nearly

(A) 10.2 kPa

(B) 10.8 kPa

(C) 11.3 kPa

(D) 12.7 kPa

4. The hydraulic radius of a 5 m deep triangular channel with a 1:1 side slope is most nearly

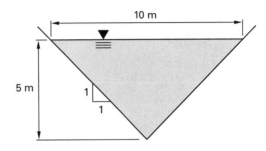

(A) 1.0 m

(B) 1.8 m

(C) 2.0 m

(D) 2.8 m

5. A pipe carrying an incompressible fluid has a diameter of 100 mm at point 1 and a diameter of 50 mm at point 2. The velocity of the fluid at point 1 is 0.3 m/s. What is most nearly the flow velocity at point 2?

(A) 0.95 m/s

(B) 1.2 m/s

(C) 2.1 m/s

(D) 3.5 m/s

SOLUTIONS

1. The Froude numbers must be equal.

$$Fr_{dam} = Fr_{model}$$

$$\frac{v_{dam}}{\sqrt{gy_{h,dam}}} = \frac{v_{model}}{\sqrt{gy_{h,model}}}$$

$$v_{dam} = v_{model}\sqrt{\frac{y_{h,dam}}{y_{h,model}}} = \left(5\ \frac{m}{s}\right)\sqrt{\frac{15}{1}}$$

$$= 19.36\ m/s \quad (19\ m/s)$$

The answer is (D).

2. The volumetric flow rate is

$$Q = \frac{10\ \dfrac{L}{s}}{1000\ \dfrac{L}{m^3}} = 0.01\ m^3/s$$

The velocity needed is

$$v = \frac{Q}{A} = \frac{0.01\ \dfrac{m^3}{s}}{0.5\ m^2} = 0.02\ m/s$$

Use Manning's equation to find the slope needed to achieve this velocity.

$$v = \left(\frac{K}{n}\right)R_H^{2/3}S^{1/2}$$

$$S = \left(\frac{vn}{KR_H^{2/3}}\right)^2$$

$$= \left(\frac{\left(0.02\ \dfrac{m}{s}\right)(0.15)}{(1.0)(0.15\ m)^{2/3}}\right)^2$$

$$= 0.0001129 \quad (1.1 \times 10^{-4})$$

The answer is (A).

3. From the continuity equation for incompressible flow,

$$A_1 v_1 = A_2 v_2$$

$$v_2 = \frac{A_1 v_1}{A_2} = \left(\frac{D_1}{D_2}\right)^2 v_1$$

$$= \left(\frac{150\ mm}{75\ mm}\right)^2 \left(1.2\ \frac{m}{s}\right)$$

$$= 4.8\ m/s$$

Use the Bernoulli equation.

$$\frac{p_2}{\rho} + \frac{v_2^2}{2} + z_2 g = \frac{p_1}{\rho} + \frac{v_1^2}{2} + z_1 g$$

$$z_1 = z_2$$

$$p_2 = 0 \quad [gage]$$

$$p_1 = \frac{\rho}{2}(v_2^2 - v_1^2)$$

$$= \left(\frac{1000\ \dfrac{kg}{m^3}}{2}\right)\left(\left(4.8\ \frac{m}{s}\right)^2 - \left(1.2\ \frac{m}{s}\right)^2\right)$$

$$= 10\,800\ Pa \quad (10.8\ kPa)$$

The answer is (B).

4. The hydraulic radius is found by dividing the cross-sectional area of the channel by the wetted perimeter. Because the sides have a 1:1 slope, the cross section of the channel is a 90°-45°-45° triangle with the vertex down. The width of the channel surface (the base of the inverted triangle) is 10 m, and the area of the triangular cross section is

$$A = \frac{bh}{2} = \frac{(10\ m)(5\ m)}{2} = 25\ m^2$$

The wetted length of each side of the channel is equal to the hypotenuse of a 90°-45°-45° triangle with sides of 5 m, and the wetted perimeter is the sum of these two wetted lengths.

$$hypotenuse = \sqrt{a^2 + b^2} = \sqrt{(5\ m)^2 + (5\ m)^2}$$

$$= 7.071\ m$$

$$wetted\ perimeter = (2)(7.071\ m) = 14.14\ m$$

The hydraulic radius is

$$R_H = \frac{cross\text{-}sectional\ area}{wetted\ perimeter} = \frac{25\ m^2}{14.14\ m}$$

$$= 1.768\ m \quad (1.8\ m)$$

The answer is (B).

5. Use the continuity equation for an incompressible fluid, and solve for the velocity at point 2.

$$A_1 v_1 = A_2 v_2$$

$$\left(\frac{\pi D_1^2}{4}\right)v_1 = \left(\frac{\pi D_2^2}{4}\right)v_2$$

$$v_2 = \left(\frac{D_1}{D_2}\right)^2 v_1 = \left(\frac{100\ mm}{50\ mm}\right)^2\left(0.3\ \frac{m}{s}\right)$$

$$= 1.2\ m/s$$

The answer is (B).

Hydraulics/
Hydrologic Sys.

11 Groundwater

PRACTICE PROBLEMS

1. Water drains at a constant rate through a saturated soil column with a diameter of 1 ft and a height of 2 ft. The hydraulic head is maintained at 5 ft at the top of the column and 0.5 ft at the bottom. After a period of 1 hr, 100 in^3 of water has drained through the column. What is most nearly the hydraulic conductivity of the soil?

(A) 3.5×10^{-6} ft/sec

(B) 4.6×10^{-6} ft/sec

(C) 7.1×10^{-6} ft/sec

(D) 9.1×10^{-6} ft/sec

2. Darcy's law is primarily associated with flow through

(A) open channels

(B) pipes

(C) pitot tubes and venturi meters

(D) porous media

3. A soil sample with a permeability of 5×10^{-6} in/sec will be tested using the pipe setup shown. The pipe's diameter is 2 in. The 10 in head differential will be maintained.

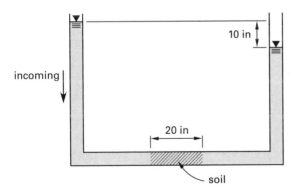

The volume of flow is most nearly

(A) 2.5×10^{-6} in^3/sec

(B) 4.9×10^{-6} in^3/sec

(C) 7.9×10^{-6} in^3/sec

(D) 3.0×10^{-5} in^3/sec

4. An aquifer has a thickness of 52 ft and a transmissivity of 650 ft^2/day. What is most nearly the hydraulic conductivity of the aquifer?

(A) 2.5 ft/day

(B) 6.3 ft/day

(C) 13 ft/day

(D) 33 ft/day

5. The results of well pumping tests from a homogeneous, unconfined aquifer are shown. At the time of the tests, the pumping had continued long enough for the well discharge to become steady.

parameter	value
pumping rate	20 gal/sec
well diameter	1.5 ft
radius of influence	900 ft
depth of aquifer at radius of influence	135 ft
drawdown in well	11 ft

The hydraulic conductivity of the aquifer is most nearly

(A) 1.2×10^{-3} ft/sec

(B) 2.1×10^{-3} ft/sec

(C) 5.0×10^{-3} ft/sec

(D) 8.7×10^{-3} ft/sec

SOLUTIONS

1. The cross-sectional area of the column is

$$A = \frac{\pi D^2}{4} = \frac{\pi(1 \text{ ft})^2}{4} = 0.785 \text{ ft}^2$$

The change in hydraulic head over the length of the soil sample is

$$\frac{dh}{dx} = \frac{0.5 \text{ ft} - 5 \text{ ft}}{2 \text{ ft}} = -2.25 \text{ ft/ft}$$

Use Darcy's law.

$$Q = -KA\left(\frac{dh}{dx}\right)$$

$$K = -\frac{Q}{A\dfrac{dh}{dx}}$$

$$= -\frac{100 \dfrac{\text{in}^3}{\text{hr}}}{(0.785 \text{ ft}^2)\left(-2.25 \dfrac{\text{ft}}{\text{ft}}\right)}$$

$$\times \left(12 \frac{\text{in}}{\text{ft}}\right)^3 \left(3600 \frac{\text{sec}}{\text{hr}}\right)$$

$$= 9.097 \times 10^{-6} \text{ ft/sec} \quad (9.1 \times 10^{-6} \text{ ft/sec})$$

The answer is (D).

2. Darcy's law is primarily associated with flow though porous media.

The answer is (D).

3. The cross-sectional area of the pipe is

$$A = \frac{\pi D^2}{4} = \frac{\pi(2 \text{ in})^2}{4} = 3.142 \text{ in}^2$$

The change in hydraulic head in the direction of flow over the length of the soil sample is

$$\frac{dh}{dx} = \frac{-10 \text{ in}}{20 \text{ in}} = -0.5 \text{ in/in}$$

Use Darcy's law.

$$Q = -KA\left(\frac{dh}{dx}\right)$$

$$= -\left(5 \times 10^{-6} \frac{\text{in}}{\text{sec}}\right)(3.142 \text{ in}^2)\left(-0.5 \frac{\text{in}}{\text{in}}\right)$$

$$= 7.854 \times 10^{-6} \text{ in}^3/\text{sec} \quad (7.9 \times 10^{-6} \text{ in}^3/\text{sec})$$

The answer is (C).

4. From the formula for transmissivity, the permeability is

$$T = KD$$

$$K = \frac{T}{D} = \frac{650 \dfrac{\text{ft}^2}{\text{day}}}{52 \text{ ft}}$$

$$= 12.5 \text{ ft/day} \quad (13 \text{ ft/day})$$

The answer is (C).

5. The radius of the well is

$$r_2 = \frac{D}{2} = \frac{1.5 \text{ ft}}{2} = 0.75 \text{ ft}$$

The thickness of the saturated aquifer at the well is

$$h_2 = D - D_{w,2} = 135 \text{ ft} - 11 \text{ ft} = 124 \text{ ft}$$

Use Dupuit's equation to find the hydraulic conductivity of the aquifer.

$$Q = \frac{\pi k\left(h_2^2 - h_1^2\right)}{\ln \dfrac{r_2}{r_1}}$$

$$k = \frac{Q \ln \dfrac{r_2}{r_1}}{\pi(h_2^2 - h_1^2)}$$

$$= \frac{\left(\dfrac{20 \dfrac{\text{gal}}{\text{sec}}}{7.48 \dfrac{\text{gal}}{\text{ft}^3}}\right) \ln \dfrac{0.75 \text{ ft}}{900 \text{ ft}}}{\pi\left((124 \text{ ft})^2 - (135 \text{ ft})^2\right)}$$

$$= 2.118 \times 10^{-3} \text{ ft/sec} \quad (2.1 \times 10^{-3} \text{ ft/sec})$$

The answer is (B).

12 Water Quality

PRACTICE PROBLEMS

1. What is most nearly the equivalent weight of aluminum sulfate, $Al_2(SO_4)_3$?

(A) 57 g/mol

(B) 110 g/mol

(C) 170 g/mol

(D) 340 g/mol

2. A water sample contains 57 mg/L of phosphate, PO_4^{---}, measured as substance. What is most nearly the concentration as $CaCO_3$?

(A) 32 mg/L as $CaCO_3$

(B) 36 mg/L as $CaCO_3$

(C) 60 mg/L as $CaCO_3$

(D) 90 mg/L as $CaCO_3$

3. The alkalinity of water was determined to be 200 mg/L as $CaCO_3$. If a nitric acid (HNO_3; molecular weight of 63 g/mol) solution with an acidity of 35 mg/L as $CaCO_3$ is available, most nearly how much acid solution would be required to completely neutralize 700 L of the water?

(A) 440 L

(B) 3200 L

(C) 4000 L

(D) 5000 L

4. A water analysis of lake water has the results shown, with all values reported as $CaCO_3$.

alkalinity	151.5 mg/L
sodium	120.0 mg/L
calcium	127.5 mg/L
iron (III)	0.107 mg/L
magnesium	43.5 mg/L
potassium	8.24 mg/L
chloride	39.5 mg/L
fluoride	1.05 mg/L
nitrate	1.06 mg/L
sulfate	106 mg/L

The water's hardness is most nearly

(A) 150 mg/L

(B) 170 mg/L

(C) 290 mg/L

(D) 300 mg/L

5. Water has a carbonate hardness of 92 mg/L as $CaCO_3$. Most nearly, what dose of lime (CaO; oxidation number 2) as substance should be added to the water to reduce the hardness to 30 mg/L as $CaCO_3$?

(A) 35 mg/L

(B) 60 mg/L

(C) 90 mg/L

(D) 180 mg/L

6. Which of the following is generally NOT attributable to hard water?

(A) scum rings in bathtubs

(B) stains on porcelain bath fixtures

(C) foam in clothes-washing equipment

(D) scale buildup in boiler tubes

7. A fate and transport study was performed on phosphorus in a small pond. Biological processes in the pond biota convert phosphorus to a nonbioavailable form at the rate of 22% per year. Recycling of sediment phosphorus by rooted plants and by anaerobic conditions in the hypolimnion converts 12% per year of the nonbioavailable phosphorus back to bioavailable forms. Runoff into the pond evaporates during the year, so no change in pond volume occurs. Which of the following actions would best reduce the phosphorus accumulation in the pond?

(A) adding chemicals to precipitate phosphorus in the pond

(B) adding chemicals to combine with phosphorus in the pond

(C) reducing use of phosphorus-based fertilizers in the surrounding fields

(D) using natural-based soaps and detergents in surrounding homes

8. Hardness in natural water is caused by the presence of which of the following?

(A) weakly acidic ions

(B) nitrites and nitrates

(C) polyvalent metallic cations

(D) colloidal solids

9. In aquatic systems, phosphorus recycling is significantly aided by which of the following?

I. algae

II. fungi

III. macrophytes

IV. phytoplankton

(A) I and II

(B) I, II, and III

(C) I, III, and IV

(D) II, III, and IV

10. Which of the following are sources of color in water?

I. copper ions

II. iron ions

III. manganese ions

IV. industrial colloidal solids

(A) IV only

(B) I and II

(C) I, II, and IV

(D) I, II, III, and IV

11. Which of the following processes is used to measure the amount of acid in water?

(A) titration

(B) filtration

(C) spectrophotometry

(D) digestion

12. Acidity in water is typically specified in terms of which of the following?

(A) H_2CO_3

(B) H^+

(C) OH^-

(D) $CaCO_3$

13. The National Primary Drinking Water Regulations apply to every public water supply serving at least how many service connections?

(A) 15

(B) 50

(C) 100

(D) 250

14. The molar nitrogen-to-phosphorus (N:P) ratio for ideal algae growth is which of the following?

(A) 10:1

(B) 12:1

(C) 16:1

(D) 18:1

15. In water, excess amounts of which of the following can contribute to methemoglobinemia, or "blue baby" syndrome?

(A) phosphorus

(B) manganese

(C) carbonates

(D) nitrates

Environmental
Engineering

SOLUTIONS

1. For aluminum sulfate, $Al_2(SO_4)_3$, the molecular weight is

$$MW_{Al_2(SO_4)_3} = 2(MW_{Al}) + 3(MW_S) + 12(MW_O)$$
$$= (2)\left(26.981 \ \frac{g}{mol}\right) + (3)\left(32.066 \ \frac{g}{mol}\right)$$
$$+ (12)\left(15.999 \ \frac{g}{mol}\right)$$
$$= 342.148 \ g/mol$$

The aluminum ion is triply charged, and there are two aluminum ions, so the number of charges involved is $(3)(2) = 6$. The equivalent weight is

$$EW_{Al_2(SO_4)_3} = \frac{MW_{Al_2(SO_4)_3}}{\text{oxidation number}}$$
$$= \frac{342.148 \ \frac{g}{mol}}{6}$$
$$= 57.025 \ g/mol \quad (57 \ g/mol)$$

The answer is (A).

2. The molecular weight of phosphate, PO_4^{---}, is

$$MW_{PO_4^{---}} = MW_P + 4(MW_O)$$
$$= 30.974 \ \frac{g}{mol} + (4)\left(15.999 \ \frac{g}{mol}\right)$$
$$= 94.97 \ g/mol$$

The ion is triply charged, so the equivalent weight is

$$EW_{PO_4^{---}} = \frac{MW_{PO_4^{---}}}{\text{oxidation number}}$$
$$= \frac{94.97 \ \frac{g}{mol}}{3}$$
$$= 31.657 \ g/mol$$

The concentration as $CaCO_3$ is

$$C_{\text{as } CaCO_3} = C_{\text{as substance}} \left(\frac{EW_{CaCO_3}}{EW_{PO_4^{---}}}\right)$$
$$= \left(57 \ \frac{mg}{L}\right)\left(\frac{50.1 \ \frac{g}{mol}}{31.657 \ \frac{g}{mol}}\right)$$
$$= 90.21 \ mg/L \quad (90 \ mg/L \text{ as } CaCO_3)$$

The answer is (D).

3. Since the alkalinity and the acidity concentrations were both given as $CaCO_3$ equivalents, the volumes can be determined without converting the concentrations. Balance the total masses of alkalinity and acidity.

$$\Delta m_{\text{alkalinity}} = \Delta m_{\text{acidity}}$$
$$V_{\text{water}}\Delta C_{\text{water,as } CaCO_3} = V_{\text{acid}}\Delta C_{\text{acid,as } CaCO_3}$$
$$V_{\text{acid}} = \frac{V_{\text{water}}\Delta C_{\text{water,as } CaCO_3}}{\Delta C_{\text{acid,as } CaCO_3}}$$
$$= \frac{(700 \ L)\left(200 \ \frac{mg}{L} - 0 \ \frac{mg}{L}\right)}{35 \ \frac{mg}{L} - 0 \ \frac{mg}{L}}$$
$$= 4000 \ L$$

The answer is (C).

4. Water hardness is determined from the polyvalent metallic cations, which are calcium (Ca^{++}), iron (Fe^{+++}), and magnesium (Mg^{++}). Sodium and potassium are singly charged and do not contribute to hardness.

$$\text{hardness} = C_{Ca^{++}} + C_{Fe^{+++}} + C_{Mg^{++}}$$
$$= 127.5 \ \frac{mg}{L} + 0.107 \ \frac{mg}{L} + 43.5 \ \frac{mg}{L}$$
$$= 171.107 \ mg/L \quad (170 \ mg/L)$$

The answer is (B).

5. The lime is being added to the water, so the volume of solution is the same as the volume of treated water. Determine the $CaCO_3$ dose.

$$V_{\text{water}}\Delta C_{\text{water,as } CaCO_3} = V_{\text{lime}}D_{\text{lime,as } CaCO_3}$$
$$D_{\text{lime,as } CaCO_3} = \frac{V_{\text{water}}\Delta C_{\text{water,as } CaCO_3}}{V_{\text{lime}}}$$
$$= \frac{(1 \ L)\left(92 \ \frac{mg}{L} - 30 \ \frac{mg}{L}\right)}{1 \ L}$$
$$= 62 \ mg/L$$

Determine the $CaCO_3$ equivalent of CaO. The molecular weight of CaO is

$$MW_{\text{lime}} = MW_{Ca} + MW_O$$
$$= 40.078 \ \frac{g}{mol} + 15.999 \ \frac{g}{mol}$$
$$= 56.077 \ g/mol$$

The calcium ion is doubly charged, so the equivalent weight of lime is

$$EW_{\text{lime}} = \frac{MW_{\text{lime}}}{\text{oxidation number}} = \frac{56.077 \ \frac{g}{mol}}{2}$$
$$= 28.039 \ g/mol$$

Environmental Engineering

The concentration as substance is

$$D_{\text{lime,as substance}} = D_{\text{lime,as CaCO}_3} \left(\frac{EW_{\text{lime}}}{EW_{\text{CaCO}_3}} \right)$$

$$= \left(62 \ \frac{\text{mg}}{\text{L}} \right) \left(\frac{28.039 \ \frac{\text{g}}{\text{mol}}}{50.1 \ \frac{\text{g}}{\text{mol}}} \right)$$

$$= 34.7 \ \text{mg/L} \quad (35 \ \text{mg/L})$$

The answer is (A).

6. Hard water interferes with foam/bubble formation, turning soap into scum.

The answer is (C).

7. To reduce the phosphorus accumulation in the pond, the arrival of additional bioavailable phosphorus must be reduced. This entails watershed management to reduce the amount of phosphorus applied as fertilizer and released through other sources. As eutrophication is a natural process accelerated by the availability of plant nutrients (nitrogen and phosphorus especially), reducing phosphorus can slow the process.

While there are chemical means to alter the forms of phosphorus to nonbioavailable states, this usually is not practical on a large scale and is potentially harmful in itself.

It is unlikely that the pond receives untreated discharge from local homes. Use of phosphate-rich detergents in the home will not affect the pond.

The answer is (C).

8. Hardness in natural water is caused by the presence of polyvalent metallic cations.

The answer is (C).

9. In aquatic systems, phosphorus recycling is significantly aided by algae, macrophytes, and phytoplankton.

The answer is (C).

10. Copper, iron, and manganese ions, and industrial colloidal solids are all sources of color in water.

The answer is (D).

11. Titration is used to measure the amount of acid in water.

The answer is (A).

12. The measurement of acidity in water is typically given in terms of the $CaCO_3$ equivalent that would neutralize the acid.

The answer is (D).

13. Every public water supply serving 15 or more service connections must meet the National Primary Drinking Water Regulations.

The answer is (A).

14. The molar N:P ratio for ideal algae growth is 16:1.

The answer is (C).

15. Excess amounts of nitrate in water can contribute to methemoglobinemia.

The answer is (D).

Environmental
Engineering

13 Water Supply Treatment and Distribution

PRACTICE PROBLEMS

1. Primary treatment of wastewater consists of

(A) chlorination

(B) removal of toxic industrial wastes

(C) regulating the flow of the wastewater through tanks of varied sizes so that the flow is a constant in secondary and tertiary treatment

(D) removing solids and particles of various sizes

2. In a packed bed operating at 25°C, a vapor contained in an air stream is dissolved in water. The Henry's law constant of the vapor is 0.10 atm·L/mol. The volumetric flow rate of air is 1 m³/s. Most nearly, what volumetric flow rate of water is required?

(A) 2 L/s

(B) 4 L/s

(C) 6 L/s

(D) 10 L/s

3. At 15°C, the Henry's law constant for ammonia is 0.62 atm. The concentration of ammonia in a water solution at this temperature is 8.1×10^{-3} mol/L. The partial pressure of ammonia vapor above the liquid phase is most nearly

(A) 9.0×10^{-5} atm

(B) 4.4×10^{-5} atm

(C) 3.8×10^{-5} atm

(D) 5.0×10^{-3} atm

4. The equilibrium equation for the dissolution of sulfur dioxide in water is

$$SO_2(g) + 2H_2O(l) \rightleftharpoons HSO_3^-(aq) + H_3O^+(aq)$$

The equilibrium constant for this equation is

$$K_{eq} = \frac{[HSO_3^-][H_3O^+]}{p_{SO_2}} = 2.1 \times 10^{-2} \ (mol/L)^2/atm$$

The variable p_{SO_2} represents the partial pressure of sulfur dioxide in the gaseous phase. The molar concentration of hydronium ions in water in equilibrium with air that contains 0.1 parts per million volume (ppmv) of sulfur dioxide is most nearly

(A) 2.1×10^{-9} mol/L

(B) 4.6×10^{-5} mol/L

(C) 5.3×10^{-5} mol/L

(D) 2.1×10^{-3} mol/L

5. Which of the following diseases should be of concern to the environmental engineer when designing, operating, or managing water supply projects?

I. acquired immunodeficiency syndrome (AIDS)

II. Rocky Mountain spotted fever

III. botulism

IV. tuberculosis

V. Legionnaires' disease

VI. hepatitis A

(A) V only

(B) V and VI

(C) I, III, and VI

(D) II, IV, and VI

6. Which of the following substances are used to remove calcium and magnesium from hard water?

I. lime

II. soda ash

III. ozone

IV. activated carbon

(A) I and II

(B) I and III

(C) II and III

(D) III and IV

7. Water is recarbonated after it has been softened to

I.   remove unwanted odors

II.   lower its pH

III. improve its taste

IV.   reduce its scale-forming potential

    (A) I and II

    (B) II and III

    (C) II and IV

    (D) III and IV

8. In addition to lime-soda treatment, what method can be used to reduce the hardness of water?

    (A) neutralizing alkalinity with hydrochloric acid

    (B) replacing calcium ions with sodium ions

    (C) adsorbing sodium and iron with granular activated carbon

    (D) flocculating the water with magnesium sulfate ($MgSO_4$)

9. Noncarbonate hardness can be removed by which of the following?

I.   heating

II.   distillation

III. precipitation softening processes

IV.   ion exchange processes

    (A) I and II

    (B) I and III

    (C) II and IV

    (D) III and IV

10. Which of the following are generally true for water treatment relative to the adsorption of a contaminate by activated carbon?

I.   The adsorption is a chemical reaction and typically irreversible.

II.   The adsorption is a physical reaction (van der Waals forces) and generally reversible.

III. Water soluble, inorganic contaminants with low molecular weights are best adsorbed by activated carbon.

IV.   The contaminant sticks to the surface of the activated carbon particles.

    (A) II and IV

    (B) III and IV

    (C) I, II, and III

    (D) I, II, III, and IV

11. 1 L of water having a contaminant concentration of 1 mg/L is to be treated with activated carbon. A residual or equilibrium contaminant concentration of 0.1 mg/L is desired. The activated carbon can adsorb 5% of its weight at the desired equilibrium concentration. Most nearly, what mass of activated carbon is required?

    (A) 0.9 mg

    (B) 18 mg

    (C) 20 mg

    (D) 29 mg

12. The electrodialysis of a 0.1 N NaCl solution is carried out in 100 cells. The flow rate of the solution is 1 L/s; the removal efficiency is 60%, and the electrical efficiency is 98%. The current required for operation is most nearly

    (A) 60 A

    (B) 200 A

    (C) 500 A

    (D) 1000 A

Environmental Engineering

SOLUTIONS

1. The primary treatment of wastewater is the removal of solids and particles.

The answer is (D).

2. The volumetric flow rate is

$$Q_W = Q_A H' = \frac{Q_A H}{RT}$$

$$= \frac{\left(1 \ \frac{\text{m}^3}{\text{s}}\right)\left(1000 \ \frac{\text{L}}{\text{m}^3}\right)\left(0.10 \ \frac{\text{atm·L}}{\text{mol}}\right)}{\left(0.08206 \ \frac{\text{atm·L}}{\text{mol·K}}\right)(25°\text{C} + 273°)}$$

$$= 4.09 \ \text{L/s} \quad (4 \ \text{L/s})$$

The answer is (B).

3. The concentration of ammonia is 8.1×10^{-3} moles per liter of solution. The solution is essentially 100% water. The molecular weight of water (H_2O) is

$$MW_{H_2O} = 2(MW_H) + MW_O$$

$$= (2)\left(1.0079 \ \frac{\text{g}}{\text{mol}}\right) + 15.999 \ \frac{\text{g}}{\text{mol}}$$

$$= 18.0148 \ \text{g/mol}$$

The number of moles of water in a liter of water is

$$n_W = \frac{m}{MW_{H_2O}} = \frac{\rho V}{MW_{H_2O}}$$

$$= \frac{\left(1000 \ \frac{\text{kg}}{\text{m}^3}\right)(1 \ \text{L})\left(1000 \ \frac{\text{g}}{\text{kg}}\right)}{\left(18.0148 \ \frac{\text{g}}{\text{mol}}\right)\left(1000 \ \frac{\text{L}}{\text{m}^3}\right)}$$

$$= 55.51 \ \text{mol}$$

The mole fraction of ammonia in the solution is

$$x = \frac{n_{\text{ammonia}}}{n_W} = \frac{8.1 \times 10^{-3} \ \text{mol}}{55.51 \ \text{mol}} = 1.459 \times 10^{-4}$$

Use Henry's law.

$$p_i = hx_i = (0.62 \ \text{atm})(1.459 \times 10^{-4})$$

$$= 9.047 \times 10^{-5} \ \text{atm} \quad (9.0 \times 10^{-5} \ \text{atm})$$

The answer is (A).

4. For mixtures of ideal gases, the volumetric fraction and the mole fraction are the same.

$$x_{SO_2} = 0.1 \times 10^{-6}$$

Use Henry's law. The partial pressure of SO_2 is

$$p_{SO_2} = x_{SO_2} p_{\text{total}}$$

$$= (0.1 \times 10^{-6})(1 \ \text{atm})$$

$$= 1.0 \times 10^{-7} \ \text{atm}$$

When SO_2 dissociates, an equal number of HSO_3^- and H_3O^+ ions are created, so

$$[HSO_3^-][H_3O^+] = [H_3O^+]^2$$

Solve the equilibrium constant equation for the $[H_3O^+]$ concentration.

$$[H_3O^+] = \sqrt{p_{SO_2} K_{\text{eq}}}$$

$$= \sqrt{(1.0 \times 10^{-7} \ \text{atm})\left(2.1 \times 10^{-2} \ \frac{\left(\frac{\text{mol}}{\text{L}}\right)^2}{\text{atm}}\right)}$$

$$= 4.6 \times 10^{-5} \ \text{mol/L}$$

The answer is (B).

5. Items I through IV are not concerns: AIDS is not a waterborne disease; Rocky Mountain spotted fever is passed to humans by ticks; botulism is foodborne or can enter the body through contaminated soil or needles; and tuberculosis is spread by inhalation of infectious droplets.

Legionnaires' disease, on the other hand, is associated with heat transfer systems, warm temperature water, and stagnant water. Hepatitis A can also be spread through contaminated water.

The answer is (B).

6. Lime and soda ash are used to remove calcium and magnesium from hard water.

The answer is (A).

7. Water is recarbonated after it has been softened to lower its pH and reduce its scale-forming potential.

The answer is (C).

8. Various ion exchange methods are used to replace multivalent ions such as calcium (Ca^{++}) and magnesium (Mg^{++}) with monovalent ions such as sodium (Na^+).

The answer is (B).

9. Noncarbonate hardness cannot be removed by heating. It can be removed by precipitation softening processes (typically, the lime-soda ash process) or by ion exchange processes using resins selective for ions causing hardness.

The answer is (D).

Environmental Engineering

10. Adsorption is a physical process wherein the contaminant adheres to the surface of an adsorbent such as granular activated charcoal. Following saturation, the adsorbent can be reactivated in a number of ways and reused. Adsorbents are particularly effective at removing large organic molecules from the processing stream.

The answer is (A).

11. Since 1 L of water is to be treated, and since the concentrations are given per liter, work on a per-liter basis. The mass of contaminant removed is

$$x_{\text{removed}} = V(C_{\text{in}} - C_{\text{out}})$$
$$= (1 \text{ L})\left(1 \frac{\text{mg}}{\text{L}} - 0.1 \frac{\text{mg}}{\text{L}}\right)$$
$$= 0.9 \text{ mg}$$

Solve the equation for mass ratio of the solid phase for the mass of adsorbent.

$$m_{\text{adsorbent}} = \frac{x_{\text{removed}}}{X} = \frac{0.9 \text{ mg}}{0.05 \dfrac{\text{g}}{\text{g}}} = 18 \text{ mg}$$

The answer is (B).

12. The required current is

$$I = (FQN/n) \times E_1/E_2$$
$$= \left(\frac{\left(96\,485 \dfrac{\text{C}}{\text{g·equivalent}}\right)\left(1 \dfrac{\text{L}}{\text{s}}\right)}{\dfrac{\times \left(0.1 \dfrac{\text{g·equivalent}}{\text{L}}\right)}{100}}\right)\left(\frac{0.60}{0.98}\right)$$

$$= 59.07 \text{ A} \quad (60 \text{ A})$$

The answer is (A).

14 Wastewater Collection and Treatment

PRACTICE PROBLEMS

1. The solids loading rate for a 30.5 m diameter clarifier with a flow rate of 5 MGD and an influent total suspended solids content of 150 mg/L is most nearly

(A) 1.1 kg/d·m^2

(B) 2.2 kg/d·m^2

(C) 3.9 kg/d·m^2

(D) 4.2 kg/d·m^2

2. What can treated wastewater be used for?

I. irrigation

II. fire fighting

III. road maintenance

IV. bathing

(A) I and II

(B) II and III

(C) I, II, and III

(D) I, II, III, and IV

3. How is most wastewater phosphorus removed?

(A) primary sedimentation

(B) flocculation and sedimentation

(C) biological processes

(D) adsorption in granular activated carbon towers

4. What is the primary factor that determines whether a stabilization pond will be aerobic or anaerobic?

(A) detention time

(B) depth

(C) surface (plan) area

(D) temperature

5. Which type of stabilization ponds are the most common for small communities?

(A) facultative ponds

(B) aerobic ponds

(C) anaerobic ponds

(D) aerated lagoons

6. Which of the following are used in the secondary treatment of wastewater?

I. biological beds

II. electrodialysis

III. rotating contactors

IV. air stripping

(A) III only

(B) I and II

(C) II and III

(D) I, II, III, and IV

SOLUTIONS

1. The solids loading rate is

$$
\begin{aligned}
\text{solids loading} \atop \text{rate} &= \frac{QX}{A} = \frac{QX}{\dfrac{\pi D^2}{4}} \\[2ex]
&= \frac{\left(5 \times 10^6 \ \dfrac{\text{gal}}{\text{d}}\right)\left(150 \ \dfrac{\text{mg}}{\text{L}}\right)}{\left(\pi \dfrac{(30.5 \ \text{m})^2}{4}\right)\left(1000 \ \dfrac{\text{mg}}{\text{g}}\right)} \\[1ex]
&\qquad \times \left(1000 \ \dfrac{\text{g}}{\text{kg}}\right) \\[1ex]
&= 3.886 \ \text{kg/d·m}^2 \quad (3.9 \ \text{kg/d·m}^2)
\end{aligned}
$$

The answer is (C).

2. Treated wastewater is generally not considered potable. Treated wastewater can be used for irrigation, firefighting, and road maintenance.

The answer is (C).

3. Most phosphorus is soluble. It must be removed through flocculation and sedimentation.

The answer is (B).

4. The average depth of an aerobic stabilization pond is about 1.3 m, so aerobic conditions can be maintained throughout through photosynthesis and mixing due to wind and convection. Anaerobic ponds are much deeper, and the biological organisms rapidly deplete the oxygen at the deeper levels.

The answer is (B).

5. Facultative ponds are the most common pond type selected for small communities.

The answer is (A).

6. Rotating contactors and biological beds are used in the secondary treatment of wastewater.

The answer is (C).

15 Activated Sludge and Sludge Processing

PRACTICE PROBLEMS

1. A 770,000 gal conventional activated sludge process has a recycle ratio of 0.25, an influent flow rate of 1 MGD, a mixed liquor volatile suspended solids (MLVSS) concentration of 700 mg/L, and an influent BOD_5 of 160 mg/L. What is most nearly the F:M ratio of the process?

(A) 0.1 lbm/day-lbm

(B) 0.2 lbm/day-lbm

(C) 0.3 lbm/day-lbm

(D) 0.4 lbm/day-lbm

2. Sludge is digested after being removed from a secondary treatment tank. During the digestion process, the gas that is produced is mostly

(A) carbon dioxide

(B) methane

(C) nitrogen oxide

(D) hydrogen sulfide

3. A 7400 ft^3 conventional activated sludge system has a recycle ratio of 0.26, an influent flow rate of 9100 ft^3/day, a mixed liquor volatile suspended solids (MLVSS) concentration of 490 mg/L, and an influent BOD_5 of 110 mg/L. What is most nearly the F:M ratio of the system?

(A) 0.028 day^{-1}

(B) 0.28 day^{-1}

(C) 0.72 day^{-1}

(D) 1.1 day^{-1}

4. Activated sludge is treated in a 1400 ft^3 aeration basin. The biomass concentration is 1800 mg/L. The waste sludge flows at a rate of 140 ft^3/day, and the waste sludge suspended solids concentration is 600 mg/L. The effluent flows at a rate of 240 ft^3/day, and the effluent suspended solids concentration is 1900 mg/L. What is most nearly the solids residence time?

(A) 1 day

(B) 3 days

(C) 5 days

(D) 8 days

5. A sample of influent water has a sludge volume index of 65 mL/g. After 30 min of settling, the sludge volume is 260 mL/L. What is most nearly the concentration of mixed liquor suspended solids (MLSS) in the sample?

(A) 2.0×10^3 mg/L

(B) 3.0×10^3 mg/L

(C) 4.0×10^3 mg/L

(D) 5.0×10^3 mg/L

6. An aerobic digester has an influent average flow rate of 110 ft^3/day and a suspended solids concentration of 12 mg/L in the reactor. The influent 5-day BOD is 310 mg/L, and the influent suspended solids concentration is 6.0 mg/L. The reaction rate constant is 0.75 day^{-1}, and the volatile fraction is 0.3. The solids resistance time (sludge age) is 3 days. The fraction of influent 5-day BOD consisting of raw sewage is 0.5. What is most nearly the volume of the digester?

(A) 1700 ft^3

(B) 2100 ft^3

(C) 2400 ft^3

(D) 2600 ft^3

7. A high-rate, two-stage anaerobic digester treats raw sludge with an input rate of 420 ft^3/day. The digester sludge accumulates at a rate of 110 ft^3/day. The sludge reacts after 10 hr, thickens after 7 hr, and is stored for 15 hr. What is most nearly the reactor volume of the first stage?

(A) 140 ft^3

(B) 180 ft^3

(C) 280 ft^3

(D) 460 ft^3

8. Activated sludge is treated in an aeration tank. The hydraulic residence time is 5 hr, and the solids residence time is 6 days. The influent BOD concentration is 170 mg/L, and the effluent BOD concentration is 5.2 mg/L. With a yield coefficient of 0.4 and a microbial death ratio of 0.1 day^{-1}, what is most nearly the biomass concentration in the tank?

(A) 1200 mg/L

(B) 2400 mg/L

(C) 3100 mg/L

(D) 4300 mg/L

9. A high-rate, two-stage anaerobic digester treats raw sludge with an input rate of 1200 ft^3/day. The digester sludge accumulates at a rate of 550 ft^3/day. The sludge reacts after 12 hr, thickens after 10 hr, and is stored for 20 hr. What is most nearly the reactor volume at the second stage?

(A) 460 ft^3

(B) 660 ft^3

(C) 820 ft^3

(D) 980 ft^3

10. A 15,000 ft^3 conventional activated sludge system has a recycle ratio of 0.38, an influent flow rate of 17,000 ft^3/day, a mixed liquor volatile suspended solids (MLVSS) concentration of 280 mg/L, and an influent BOD_5 of 60 mg/L. What is most nearly the recycle rate of the system?

(A) 2800 ft^3/day

(B) 3600 ft^3/day

(C) 5300 ft^3/day

(D) 6500 ft^3/day

SOLUTIONS

1. The food-to-microorganism ratio is

$$\text{F:M} = Q_0 S_0 / (\text{Vol}\, X_A)$$

$$= \frac{\left(1 \times 10^6 \ \frac{\text{gal}}{\text{day}}\right)\left(160 \ \frac{\text{mg}}{\text{L}}\right)}{(770{,}000 \ \text{gal})\left(700 \ \frac{\text{mg}}{\text{L}}\right)}$$

$$= 0.297 \ \text{lbm/day-lbm} \quad (0.3 \ \text{lbm/day-lbm})$$

The answer is (C).

2. The gas produced is primarily methane.

The answer is (B).

3. The food-to-microorganism ratio is

$$\text{F:M} = Q_0 S_0 / (\text{Vol}\, X_A)$$

$$= \frac{\left(9100 \ \frac{\text{ft}^3}{\text{day}}\right)\left(110 \ \frac{\text{mg}}{\text{L}}\right)}{(7400 \ \text{ft}^3)\left(490 \ \frac{\text{mg}}{\text{L}}\right)}$$

$$= 0.28 \ \text{day}^{-1}$$

The answer is (B).

4. The solids residence time is

$$\theta_c = \frac{V(X_A)}{Q_w X_w + Q_e X_e}$$

$$= \frac{(1400 \ \text{ft}^3)\left(1800 \ \frac{\text{mg}}{\text{L}}\right)}{\left(140 \ \frac{\text{ft}^3}{\text{day}}\right)\left(600 \ \frac{\text{mg}}{\text{L}}\right)}$$

$$+ \left(240 \ \frac{\text{ft}^3}{\text{day}}\right)\left(1900 \ \frac{\text{mg}}{\text{L}}\right)$$

$$= 4.667 \ \text{days} \quad (5 \ \text{days})$$

The answer is (C).

5. Calculate the amount of MLSS.

$$\text{SVI} = \frac{\text{sludge volume after settling (mL/L)} * 1000}{\text{MLSS (mg/L)}}$$

$$\text{MLSS} = \frac{\text{sludge volume after settling (mL/L)} * 1000}{\text{SVI}}$$

$$= \frac{\left(260 \ \frac{\text{mL}}{\text{L}}\right)\left(1000 \ \frac{\text{mg}}{\text{g}}\right)}{65 \ \frac{\text{mL}}{\text{g}}}$$

$$= 4.0 \times 10^3 \ \text{mg/L}$$

The answer is (C).

6. The volume of the aerobic digester is

$$V = \frac{Q_i(X_i + FS_i)}{X_d(k_d P_v + 1/\theta_c)}$$

$$= \frac{\left(110 \ \frac{\text{ft}^3}{\text{day}}\right)\left(\left(6.0 \ \frac{\text{mg}}{\text{L}}\right) + (0.5)\left(310 \ \frac{\text{mg}}{\text{L}}\right)\right)}{\left(12 \ \frac{\text{mg}}{\text{L}}\right)\left((0.75 \ \text{day}^{-1})(0.3) + \left(\frac{1}{3 \ \text{days}}\right)\right)}$$

$$= 2643 \ \text{ft}^3 \quad (2600 \ \text{ft}^3)$$

The answer is (D).

7. The reactor volume of the first stage is

$$\text{reactor volume} = V_1 t_r$$

$$= \frac{\left(420 \ \frac{\text{ft}^3}{\text{day}}\right)(10 \ \text{hr})}{24 \ \frac{\text{hr}}{\text{day}}}$$

$$= 175 \ \text{ft}^3 \quad (180 \ \text{ft}^3)$$

The answer is (B).

8. The tank concentration is

$$X_A = \frac{\theta_c Y(S_0 - S_e)}{\theta(1 + k_d \theta_c)}$$

$$= \frac{(6 \ \text{days})(0.4)\left(170 \ \frac{\text{mg}}{\text{L}} - 5.2 \ \frac{\text{mg}}{\text{L}}\right)\left(24 \ \frac{\text{hr}}{\text{day}}\right)}{(5 \ \text{hr})\left(1 + (0.1 \ \text{day}^{-1})(6 \ \text{days})\right)}$$

$$= 1187 \ \text{mg/L} \quad (1200 \ \text{mg/L})$$

The answer is (A).

9. The reactor volume is

$$\text{reactor volume} = \frac{V_1 + V_2}{2} t_t + V_2 t_s$$

$$= \frac{\left(\dfrac{1200 \ \frac{\text{ft}^3}{\text{day}} + 550 \ \frac{\text{ft}^3}{\text{day}}}{2}\right)(10 \ \text{hr}) + \left(550 \ \frac{\text{ft}^3}{\text{day}}\right)(20 \ \text{hr})}{24 \ \frac{\text{hr}}{\text{day}}}$$

$$= 822.9 \ \text{ft}^3 \quad (820 \ \text{ft}^3)$$

The answer is (C).

10. The recycle flow rate is

$$Q_R = Q_0 R$$

$$= \left(17,000 \ \frac{\text{ft}^3}{\text{day}}\right)(0.38)$$

$$= 6460 \ \text{ft}^3/\text{day} \quad (6500 \ \text{ft}^3/\text{day})$$

The answer is (D).

Environmental
Engineering

16 Air Quality

PRACTICE PROBLEMS

1. Which of the following pollutants is NOT a criteria pollutant as defined by the National Ambient Air Quality Standards (NAAQS)?

(A) carbon dioxide

(B) lead

(C) nitrogen dioxide

(D) dust

2. Carbon monoxide has been measured in an automotive repair garage at 9 parts per million by volume (ppmv). Most nearly, what is the observed concentration per cubic meter?

(A) 11 mg

(B) 24 mg

(C) 52 mg

(D) 94 mg

3. The abbreviation VOC stands for

(A) volatile organic carbon

(B) various ocean chemicals

(C) virtual objectionable chemicals

(D) volatile organic compound

4. Which of the following equations represents the formation of acid rain?

(A) $S + O_3 + H_2O \rightarrow H_2SO_4$

(B) $SO + O_2 + H_2O \rightarrow H_2SO_4$

(C) $SO_2 + H_2O \rightarrow H_2SO_3$

(D) $SO_3 + H_2O \rightarrow H_2SO_4$

5. Incomplete combustion of fossil fuels typically results in the selective production of

(A) ozone

(B) smog

(C) carbon monoxide

(D) fly ash

6. What is the minimum size of particles generally considered to be inhalable hazards?

(A) 2.5 μm

(B) 10 μm

(C) 50 μm

(D) 250 μm

7. Which of the following devices is the LEAST effective at removing fly ash particles with a diameter of 1 μm?

(A) electrostatic precipitator

(B) fabric baghouse

(C) venturi water scrubber

(D) air cyclone

SOLUTIONS

1. The criteria pollutants are carbon monoxide, lead, nitrogen dioxide, coarse and fine particles, ozone, and sulfur dioxide.

The answer is (A).

2. Since the carbon monoxide concentration as a mass per cubic meter is desired, calculate the mass of carbon monoxide in one cubic meter of air. The molecular weight of carbon monoxide is

$$\text{MW} = 12.011 \ \frac{\text{g}}{\text{mol}} + 15.999 \ \frac{\text{g}}{\text{mol}} = 28.01 \ \text{g/mol}$$

The volume of carbon monoxide is $9 \times 10^{-6} \ \text{m}^3$, and using typical atmospheric conditions (1 atm and $16°C$), the mass of this volume is

$$m = \frac{pV}{\left(\dfrac{R}{\text{MW}}\right)T}$$

$$= \frac{\begin{array}{c}(1 \ \text{atm})(101.3 \ \text{kPa})(9 \times 10^{-6} \ \text{m}^3) \\ \times \left(1000 \ \dfrac{\text{mg}}{\text{g}}\right)\left(1000 \ \dfrac{\text{mol}}{\text{kmol}}\right)\end{array}}{\left(\dfrac{8.314 \ \dfrac{\text{kPa·m}^3}{\text{kmol·K}}}{28.01 \ \dfrac{\text{g}}{\text{mol}}}\right)(18°C + 273°)}$$

$$= 10.55 \ \text{mg} \quad (11 \ \text{mg})$$

The answer is (A).

3. VOC is an abbreviation for volatile organic compound.

The answer is (D).

4. $SO_3 + H_2O \rightarrow H_2SO_4$ represents the formation of acid rain.

The answer is (D).

5. Incomplete combustion is caused by a deficiency of oxygen during combustion. Carbon burns to CO because there is inadequate oxygen to produce CO_2.

The answer is (C).

6. Thoracic particles are generally less than 10 μm in size and are designated PM_{10}.

The answer is (B).

7. Cyclones rely on gravitational and centrifugal forces to collect particles. Small particles have small masses, rendering force-dependent collection methods less effective.

The answer is (D).

17 Soil Properties and Testing

PRACTICE PROBLEMS

1. An undisturbed sample of clay has a wet mass of 100 kg, a dry mass of 93 kg, and a total volume of 0.0491 m^3. The solids have a specific gravity of 2.65. The void ratio is most nearly

(A) 0.31

(B) 0.40

(C) 0.61

(D) 1.0

2. A saturated sample of undisturbed clay has a wet mass of 318 kg and a dry mass of 204 kg. The total volume of the sample is 0.193 m^3. Most nearly, what is the specific gravity of the soil solids?

(A) 2.4

(B) 2.6

(C) 2.7

(D) 2.9

3. A soil's grain-size distribution curve is as shown. The effective grain size is 0.19 mm, and D_{60} is 0.49 mm.

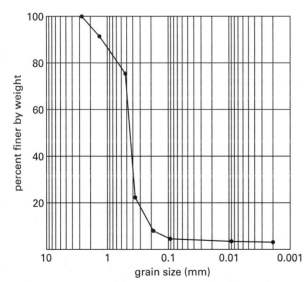

The coefficient of gradation is most nearly

(A) 0.17

(B) 0.44

(C) 1.6

(D) 3.0

4. A soil has the following characteristics.

percentage fines, F	69%
liquid limit	72
plastic limit	48

Using the Unified Soil Classification System, what is the classification of the soil?

(A) GW

(B) ML

(C) MH

(D) CH

5. A sample of soil has the following characteristics.

% passing no. 40 screen	95
% passing no. 200 screen	57
liquid limit	37
plastic limit	18

What is the AASHTO group index number?

(A) 5

(B) 6

(C) 7

(D) 8

6. The lines in a graphical seepage flow net intersect to form a pattern of

(A) triangles

(B) trapezoids

(C) squares

(D) rectangles

7. A flow net is constructed of

(A) isobars and aquicludes

(B) isobars and streamlines

(C) streamlines and isochrones

(D) isogons and isopleths

8. Granular soils tend to densify and consolidate if subjected to vibrations and seismic shaking. When saturated granular soils undergo seismic vibrations, they become subject to cyclic shear deformation, which causes an increase in pore water pressure. The result is usually a severe reduction in effective stress. This hazardous soil condition is called

(A) consolidation

(B) liquefaction

(C) heave

(D) boiling

9. The specific gravity of the soil solids in a given sample is 3.11. The porosity of the soil is 27%. What is most nearly the effective unit weight of the soil sample?

(A) 96 lbf/ft^3

(B) 110 lbf/ft^3

(C) 130 lbf/ft^3

(D) 160 lbf/ft^3

10. The total unit weight of a given soil is 120 lbf/ft^3. The soil has a porosity of 0.26 and is 100% saturated. The dry unit weight of the soil is most nearly

(A) 58 lbf/ft^3

(B) 86 lbf/ft^3

(C) 100 lbf/ft^3

(D) 120 lbf/ft^3

11. A 1 L field sample contains soil with an initial weight of 30 N. The soil is dried completely in an oven at 100°C. The weight of the soil after drying is 22 N. The maximum volume of the dry soil sample is 1.5 L, and the minimum volume is 0.60 L. The relative density of the soil is most nearly

(A) 37%

(B) 56%

(C) 62%

(D) 75%

12. An undisturbed 10 cm wide × 10 cm long × 5 cm high sample of clay has a dry weight of 32 kg. The dried soil sample is compacted into a minimum volume, a rectangular prism 8.5 cm wide × 9 cm high × 3.5 cm high. What is most nearly the relative compaction of the soil sample?

(A) 33%

(B) 45%

(C) 49%

(D) 54%

13. A soil sample has the particle size distribution shown.

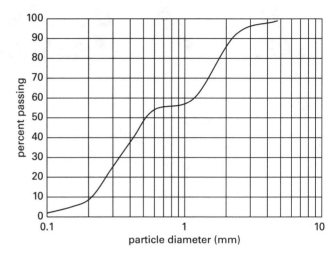

What is most nearly the coefficient of uniformity of the soil?

(A) 0.4

(B) 0.8

(C) 3

(D) 6

14. A soil has a void ratio of 0.41. The unit weight of the solids is 33 kN/m^3. What is most nearly the saturated unit weight of the soil?

(A) 26 kN/m^3

(B) 29 kN/m^3

(C) 33 kN/m^3

(D) 37 kN/m^3

15. The water content of a saturated soil sample is 26%, and the void ratio is 0.57. The unit weight of the solids is most nearly

(A) 4.5 kN/m^3

(B) 9.8 kN/m^3

(C) 22 kN/m^3

(D) 28 kN/m^3

16. An undisturbed soil sample with a volume of 0.30 ft^3 is weighed in a 0.2 lbf pan. The combined weight of the soil sample and the pan is 40.5 lbf. The soil is then completely dried in an oven at 100°C. After drying, the combined weight of the soil and the pan is 35.2 lbf. The unit weight of the soil solids is 194 lbf/ft^3. What is most nearly the volume of the air in the original soil sample?

(A) 0.010 ft^3

(B) 0.035 ft^3

(C) 0.10 ft^3

(D) 0.21 ft^3

17. A 0.10 m³ saturated soil sample is weighed in a 34 kg pan. The combined weight of the soil and the pan is 450 kg. The soil sample is completely dried in an oven. After drying, the combined weight of the soil sample and the pan is 385 kg. What is most nearly the unit weight of the saturated soil sample?

(A) 27 kN/m³

(B) 34 kN/m³

(C) 41 kN/m³

(D) 44 kN/m³

18. The results of a sieve analysis are shown.

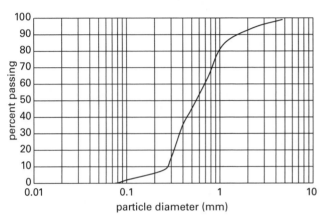

Most nearly, what is the coefficient of curvature of the soil sample?

(A) 0.3

(B) 0.7

(C) 2

(D) 3

SOLUTIONS

1. Find the volume of water in the sample. If the wet mass is 100 kg and the dry mass is 93 kg, there are 7 kg of water in the sample. The standard density of water is 1000 kg/m³, so the volume of 7 kg of water is

$$V_W = \frac{m_W}{\rho_W} = \frac{7 \text{ kg}}{1000 \ \dfrac{\text{kg}}{\text{m}^3}}$$

$$= 0.007 \text{ m}^3$$

Solve for the volume of solids using the equation for specific gravity.

$$G_S = \frac{\dfrac{m_S}{V_S}}{\rho_W}$$

$$V_S = \frac{m_S}{G_S \rho_W}$$

$$= \frac{93 \text{ kg}}{(2.65)\left(1000 \ \dfrac{\text{kg}}{\text{m}^3}\right)}$$

$$= 0.0351 \text{ m}^3$$

The volume of air in the sample is

$$V_A = V_{\text{total}} - V_W - V_S$$

$$= 0.0491 \text{ m}^3 - 0.007 \text{ m}^3 - 0.0351 \text{ m}^3$$

$$= 0.007 \text{ m}^3$$

The volume of voids is

$$V_V = V_A + V_W = 0.007 \text{ m}^3 + 0.007 \text{ m}^3$$

$$= 0.014 \text{ m}^3$$

The void ratio is

$$e = \frac{V_V}{V_S} = \frac{0.014 \text{ m}^3}{0.0351 \text{ m}^3}$$

$$= 0.399 \quad (0.40)$$

The answer is (B).

2. If the sample is saturated, there is no air in the sample. The mass of water in the sample is

$$m_W = 318 \text{ kg} - 204 \text{ kg} = 114 \text{ kg}$$

The density of water is 1000 kg/m³, so the volume of water in the sample is

$$V_W = \frac{m_W}{\rho} = \frac{114 \text{ kg}}{1000 \ \dfrac{\text{kg}}{\text{m}^3}} = 0.114 \text{ m}^3$$

Geotechnical Engineering

The volume of solids in the sample is

$$V_S = V - V_W = 0.193 \text{ m}^3 - 0.114 \text{ m}^3 = 0.079 \text{ m}^3$$

The specific gravity of the solids is

$$G_S = \frac{\dfrac{W_S}{V_S}}{\rho_W} = \frac{\dfrac{204 \text{ kg}}{0.079 \text{ m}^3}}{1000 \dfrac{\text{kg}}{\text{m}^3}}$$

$$= 2.58 \quad (2.6)$$

The answer is (B).

3. As read from the distribution curve, $D_{30} = 0.39$ mm. The coefficient of gradation is

$$C_C = \frac{D_{30}^2}{D_{10} D_{60}}$$

$$= \frac{(0.39 \text{ mm})^2}{(0.19 \text{ mm})(0.49 \text{ mm})}$$

$$= 1.63 \quad (1.6)$$

The answer is (C).

4. When F is 69%, the soil is first classified as fine-grained. From the plasticity chart, for a liquid limit of 72 and a plasticity index of 48, the soil is classified as CH.

The answer is (D).

5. The plasticity index is

$$\text{PI} = \text{LL} - \text{PL} = 37 - 18 = 19$$

The group index is

$$\text{GI} = (F - 35)(0.2 + 0.005(\text{LL} - 40))$$

$$+ 0.01(F - 15)(\text{PI} - 10)$$

$$= (57 - 35)\big(0.2 + (0.005)(37 - 40)\big)$$

$$+ (0.01)(57 - 15)(19 - 10)$$

$$= 7.85 \quad (8)$$

The answer is (D).

6. The intersections of streamlines and equipotential lines must be perpendicular and form a grid of four-sided shapes. These shapes can be referred to as curvilinear squares. The sides of the square may be curved, but perpendicular intersections are maintained.

The answer is (C).

7. A flow net uses streamlines (flow lines) and equipotential lines (isobars) to show pressure and flow path graphically. Equipotential lines represent levels of constant hydraulic head, and streamlines represent paths of flow.

The answer is (B).

8. The occurrence of liquefaction can result in loss of bearing capacity and immediate settlement of foundation elements. Excessive settlement and foundation movement can cause severe structural damage and possibly life-threatening destruction of buildings.

The answer is (B).

9. Rearrange the equation for porosity to calculate the void ratio of the soil sample.

$$n = \frac{e}{1 + e}$$

$$e = \frac{n}{1 - n}$$

$$= \frac{0.27}{1 - 0.27}$$

$$= 0.37$$

Calculate the saturated unit weight of the soil sample.

$$\gamma_{\text{sat}} = \frac{(G_S + e)\gamma_W}{1 + e}$$

$$= \frac{(3.11 + 0.37)\left(62.4 \dfrac{\text{lbf}}{\text{ft}^3}\right)}{1 + 0.37}$$

$$= 158.5 \text{ lbf/ft}^3$$

The effective unit weight of the soil sample is

$$\gamma' = \gamma_{\text{sat}} - \gamma_W$$

$$= 158.5 \frac{\text{lbf}}{\text{ft}^3} - 62.4 \frac{\text{lbf}}{\text{ft}^3}$$

$$= 96 \text{ lbf/ft}^3$$

The answer is (A).

10. The total unit weight of the soil is the combined weight of the soil and the water. For a saturated soil, the water completely fills the voids. The weight of the water is the unit weight of water multiplied by the volume of the voids.

$$\gamma_{\text{sat}} = \frac{W}{V} = \frac{W_S + W_W}{V} = \frac{W_S}{V} + \frac{W_W}{V}$$

$$= \frac{W_S}{V} + \frac{\gamma_W V_V}{V}$$

$$= \gamma_D + \gamma_W n$$

Geotechnical Engineering

The dry unit weight of the soil is

$$\gamma_D = \gamma_{\text{sat}} - \gamma_W n$$

$$= 120 \ \frac{\text{lbf}}{\text{ft}^3} - \left(62.4 \ \frac{\text{lbf}}{\text{ft}^3}\right)(0.26)$$

$$= 103.8 \ \text{lbf/ft}^3 \quad (100 \ \text{lbf/ft}^3)$$

The answer is (C).

11. Calculate the dry unit weight, maximum dry unit weight, and minimum dry unit weight of the sample.

$$\gamma_{d\text{-field}} = \frac{W_S}{V} = \frac{22 \ \text{N}}{1 \ \text{L}}$$

$$= 22 \ \text{N/L}$$

$$\gamma_{d\text{-min}} = \frac{W_S}{V} = \frac{22 \ \text{N}}{1.5 \ \text{L}}$$

$$= 14.67 \ \text{N/L}$$

$$\gamma_{d\text{-max}} = \frac{W_S}{V} = \frac{22 \ \text{N}}{0.60 \ \text{L}}$$

$$= 36.67 \ \text{N/L}$$

The relative density of the soil sample is

$$D_r = \frac{(\gamma_{d\text{-field}} - \gamma_{d\text{-min}})(\gamma_{d\text{-max}})}{(\gamma_{d\text{-max}} - \gamma_{d\text{-min}})(\gamma_{d\text{-field}})} \times 100\%$$

$$= \frac{\left(22 \ \dfrac{\text{N}}{\text{L}} - 14.67 \ \dfrac{\text{N}}{\text{L}}\right)\left(36.67 \ \dfrac{\text{N}}{\text{L}}\right)}{\left(36.67 \ \dfrac{\text{N}}{\text{L}} - 14.67 \ \dfrac{\text{N}}{\text{L}}\right)\left(22 \ \dfrac{\text{N}}{\text{L}}\right)} \times 100\%$$

$$= 56\%$$

The answer is (B).

12. Substitute the equation for dry unit weight into the equation for relative compaction of the soil sample.

$$\gamma_D = \frac{W_S}{V}$$

$$\text{RC} = \frac{\gamma_{d\text{-field}}}{\gamma_{d\text{-max}}} \times 100\%$$

$$= \frac{\left(\dfrac{W_S}{V}\right)_{\text{field}}}{\left(\dfrac{W_S}{V}\right)_{\text{max}}} \times 100\%$$

The volume in the field is the initial volume of the soil sample, and the maximum density occurs at the minimum volume. Calculate the relative compaction.

$$\text{RC} = \frac{\dfrac{W_S}{V_0}}{\dfrac{W_S}{V_{\text{min}}}} \times 100\% = \frac{V_{\text{min}}}{V_0} \times 100\%$$

$$= \frac{(8.5 \ \text{cm})(9 \ \text{cm})(3.5 \ \text{cm})}{(10 \ \text{cm})(10 \ \text{cm})(5 \ \text{cm})} \times 100\%$$

$$= 54\%$$

The answer is (D).

13. The diameter at which 10% of the particles are finer is 0.2 mm, and the diameter at which 60% of the particles are finer is 1.2 mm. Calculate the coefficient of uniformity.

$$C_U = \frac{D_{60}}{D_{10}} = \frac{1.2 \ \text{mm}}{0.2 \ \text{mm}}$$

$$= 6$$

The answer is (D).

14. Substitute the equation for unit weight of the solids into the equation for the saturated unit weight of the soil.

$$G_S = \frac{\dfrac{W_S}{V_S}}{\gamma_W}$$

$$\gamma_S = \frac{W_S}{V_S} = G_S \gamma_W$$

$$\gamma_{\text{sat}} = \frac{(G_S + e)\gamma_W}{1 + e} = \frac{G_S \gamma_W + e\gamma_W}{1 + e}$$

$$= \frac{\gamma_S + e\gamma_W}{1 + e}$$

$$= \frac{33 \ \dfrac{\text{kN}}{\text{m}^3} + (0.41)\left(9.81 \ \dfrac{\text{kN}}{\text{m}^3}\right)}{1 + 0.41}$$

$$= 26 \ \text{kN/m}^3$$

The answer is (A).

15. For a saturated soil, the volume of voids is the volume of the water.

$$\omega = \frac{W_W}{W_S}$$

$$e = \frac{V_V}{V_S} = \frac{V_W}{V_S}$$

The unit weight of the solids is

$$\frac{\omega}{e} = \frac{\dfrac{W_W}{W_S}}{\dfrac{V_W}{V_S}} = \frac{\dfrac{W_W}{V_W}}{\dfrac{W_S}{V_S}} = \frac{\gamma_W}{\gamma_S}$$

$$\gamma_S = \frac{\gamma_W e}{\omega}$$

$$= \frac{\left(9.81 \ \dfrac{\text{kN}}{\text{m}^3}\right)(0.57)}{0.26}$$

$$= 22 \ \text{kN/m}^3$$

The answer is (C).

16. The weight of the water in the sample is

$$W_W = W_{\text{pan and soil,initial}} - W_{\text{pan and soil,final}}$$

$$= 40.5 \ \text{lbf} - 35.2 \ \text{lbf}$$

$$= 5.3 \ \text{lbf}$$

The weight of the solids in the sample is

$$W_S = W_{\text{pan and soil,final}} - W_{\text{pan}}$$

$$= 35.2 \ \text{lbf} - 0.2 \ \text{lbf}$$

$$= 35.0 \ \text{lbf}$$

Calculate the volume of water in the sample.

$$V_W = \frac{W_W}{\gamma_W} = \frac{5.3 \ \text{lbf}}{62.4 \ \dfrac{\text{lbf}}{\text{ft}^3}}$$

$$= 0.085 \ \text{ft}^3$$

Calculate the volume of solids in the sample.

$$V_S = \frac{W_S}{\gamma_S} = \frac{35.0 \ \text{lbf}}{194 \ \dfrac{\text{lbf}}{\text{ft}^3}}$$

$$= 0.180 \ \text{ft}^3$$

The volume of air is

$$V_{\text{total}} = V_A + V_W + V_S$$

$$V_A = V_{\text{total}} - V_W - V_S$$

$$= 0.30 \ \text{ft}^3 - 0.085 \ \text{ft}^3 - 0.180 \ \text{ft}^3$$

$$= 0.035 \ \text{ft}^3$$

The answer is (B).

17. The unit weight of the saturated soil sample is the combined weight of the soil and water divided by the volume of the sample. Calculate the weight of the saturated soil sample.

$$W_{\text{sat}} = W_{\text{sat,soil and pan}} - W_{\text{pan}}$$

$$= \frac{(450 \ \text{kg} - 34 \ \text{kg})\left(9.81 \ \dfrac{\text{N}}{\text{kg}}\right)}{1000 \ \dfrac{\text{N}}{\text{kN}}}$$

$$= 4.08 \ \text{kN}$$

The saturated unit weight is

$$\gamma_{\text{sat}} = \frac{W_{\text{sat}}}{V} = \frac{4.08 \ \text{kN}}{0.10 \ \text{m}^3}$$

$$= 41 \ \text{kN/m}^3$$

The answer is (C).

18. The diameter at which 10% of the particles are finer is 0.28 mm. The diameter at which 30% of the particles are finer is 0.37 mm. The diameter at which 60% of the particles are finer is 0.7 mm. Calculate the coefficient of curvature.

$$C_C = \frac{D_{30}^2}{D_{10} D_{60}} = \frac{(0.37 \ \text{mm})^2}{(0.28 \ \text{mm})(0.7 \ \text{mm})}$$

$$= 0.7$$

The answer is (B).

18 Foundations

PRACTICE PROBLEMS

1. A 1 m wide continuous footing is designed to support an axial column load of 250 kN per meter of wall length. The footing is placed 0.5 m into a soil with a cohesion of 25 kPa and an angle of internal friction of 5°. Most nearly, what is the value of the Terzaghi bearing capacity factor, N_γ?

(A) 0.5

(B) 1.1

(C) 1.6

(D) 7.3

2. A concentrated vertical load of 6000 lbf is applied at the ground surface. Most nearly, what is the increase in vertical pressure 3.5 ft below the surface and 4 ft away from a point directly below the concentrated load?

(A) 29 lbf/ft^2

(B) 36 lbf/ft^2

(C) 41 lbf/ft^2

(D) 58 lbf/ft^2

3. A 10.75 cm diameter steel pile is driven 25 m into stiff, insensitive clay. The pile has an average friction capacity of 120 kPa and an ultimate tip end-bearing capacity of 7.4 kPa. The design parameters call for a factor of safety of 3. Most nearly, what is the allowable bearing capacity of the pile?

(A) 43 kN

(B) 120 kN

(C) 260 kN

(D) 340 kN

4. A 2 m wide continuous wall footing is designed to support an axial column load of 600 kN (per meter of wall length) and a moment of 250 kN·m (per meter of wall length), as shown. The footing is placed 1 m into a sandy soil with a density of 2000 kg/m^3, a cohesion of 0.5 Pa, and an angle of internal friction of 30°.

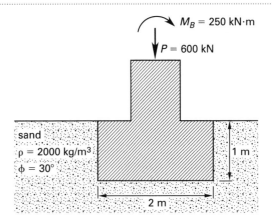

Most nearly, what is the ultimate bearing capacity per meter of footing length?

(A) 84 kPa

(B) 210 kPa

(C) 430 kPa

(D) 830 kPa

5. A 3 m wide continuous footing is designed to support an axial column load of 2000 kN per meter of wall length. The footing is placed 1.8 m into a soil with a density of 2500 kg/m^3, a cohesion of 20 kPa, and an angle of internal friction of 10°. Most nearly, what is the value of the Terzaghi bearing capacity factor, N_q?

(A) 1.2

(B) 2.7

(C) 9.6

(D) 11

SOLUTIONS

1. From a table of Terzaghi bearing capacity factors, for $\phi = 5°$, $N_\gamma = 0.5$.

The answer is (A).

2. The increase in vertical pressure is

$$\sigma_z = \frac{3Pz^3}{2\pi(r^2 + z^2)^{5/2}}$$

$$= \left(\frac{(3)(6000 \text{ lbf})}{2\pi}\right)\left(\frac{(3.5 \text{ ft})^3}{\left((4 \text{ ft})^2 + (3.5 \text{ ft})^2\right)^{5/2}}\right)$$

$$= 28.96 \text{ lbf/ft}^2 \quad (29 \text{ lbf/ft}^2)$$

The answer is (A).

3. The capacity of a pile is the sum of the skin friction capacity and the tip end-bearing capacity.

$$A_{\text{surface}} = CL = \pi DL$$

$$= \frac{\pi(10.75 \text{ cm})(25 \text{ m})}{100 \frac{\text{cm}}{\text{m}}}$$

$$= 8.44 \text{ m}^2$$

The skin friction capacity is

$$q_{\text{skin}} = A_{\text{surface}} S_{s,\text{skin}}$$

$$= (8.44 \text{ m}^2)(120 \text{ kPa})$$

$$= 1013 \text{ kN}$$

The tip area of the pile is

$$A_{\text{tip}} = \frac{\pi D^2}{4} = \frac{\pi(10.75 \text{ cm})^2}{(4)\left(100 \frac{\text{cm}}{\text{m}}\right)^2}$$

$$= 0.009 \text{ m}^2$$

The tip end-bearing capacity is

$$q_{\text{tip}} = A_{\text{tip}} q_{u,\text{tip}} = (0.009 \text{ m}^2)(7.4 \text{ kPa})$$

$$= 0.067 \text{ kN} \quad [\approx 0. \text{ Use } 0 \text{ kN.}]$$

Considering the factor of safety, the pile's capacity is

$$q_a = \frac{q_{\text{skin}} + q_{\text{tip}}}{\text{FS}} = \frac{1013 \text{ kN} + 0 \text{ kN}}{3}$$

$$= 337.7 \text{ kN} \quad (340 \text{ kN})$$

The answer is (D).

4. From a table of Terzaghi bearing capacity factors, for $\phi = 30°$, $N_c = 37.2$, $N_q = 22.5$, and $N_\gamma = 19.7$.

The ultimate bearing capacity is

$$q_u = cN_c + \gamma' D_f N_q + \tfrac{1}{2}\gamma' B N_\gamma$$

$$= (0.5 \text{ Pa})(37.2)$$

$$+ \left(2000 \frac{\text{kg}}{\text{m}^3}\right)\left(9.81 \frac{\text{m}}{\text{s}^2}\right)(1 \text{ m})(22.5)$$

$$+ (\tfrac{1}{2})\left(2000 \frac{\text{kg}}{\text{m}^3}\right)\left(9.81 \frac{\text{m}}{\text{s}^2}\right)(2 \text{ m})(19.7)$$

$$= 827\,983 \text{ Pa} \quad (830 \text{ kPa})$$

The answer is (D).

5. From a table of Terzaghi bearing capacity factors, for $\phi = 10°$, $N_q = 2.7$.

The answer is (B).

19 Rigid Retaining Walls

PRACTICE PROBLEMS

1. A soil has an angle of internal friction of 25°. What is most nearly the Rankine active earth pressure coefficient?

(A) 0.34

(B) 0.41

(C) 0.52

(D) 0.58

2. A soil has an angle of internal friction of 25°. What is most nearly the Rankine passive earth pressure coefficient?

(A) 0.59

(B) 1.6

(C) 2.5

(D) 4.1

3. A retaining wall supports soil with a vertical height of 2 m. The soil has an angle of internal friction of 32° and a specific (unit) weight of 25 kN/m³. Most nearly, what is the active lateral soil resultant?

(A) 5.0 kN/m

(B) 15 kN/m

(C) 46 kN/m

(D) 92 kN/m

4. An excavated slope in uniform soil is shown. The specific weight is 14 kN/m³, the cohesion is 15 kPa, and the angle of internal friction is 15°.

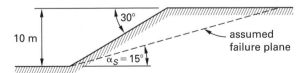

What is most nearly the available shearing resistance along the assumed failure plane?

(A) 550 kN/m

(B) 600 kN/m

(C) 680 kN/m

(D) 940 kN/m

5. A uniform soil slope has a planar slip surface length of 100 m. The soil's cohesion is 5 kPa, and the angle of internal friction is 40°. The angle that the assumed failure plane makes with respect to the horizontal is 25°. The weight of the soil mound above the assumed failure plane is 2000 kN/m. What is most nearly the mobilized shear force along the assumed failure plane?

(A) 500 kN/m

(B) 850 kN/m

(C) 1500 kN/m

(D) 2000 kN/m

6. An excavation is made in uniform soil. The available shearing resistance along an assumed slip surface is 1350 kN/m. The mobilized shear force along the slip surface is 465 kN/m. The slip surface makes an angle of 45° with respect to the horizontal. What is most nearly the factor of safety against slope instability?

(A) 1.5

(B) 2.1

(C) 2.9

(D) 4.1

7. A retaining wall extends 3 m from the top of bedrock to the ground surface. The soil is cohesionless and has the properties shown.

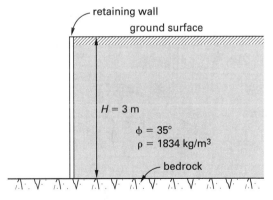

Using Rankine theory, the total active earth pressure per unit width of retaining wall is most nearly

(A) 15 kN/m

(B) 22 kN/m

(C) 44 kN/m

(D) 82 kN/m

SOLUTIONS

1. The active lateral earth pressure coefficient is

$$K_A = \tan^2\left(45° - \frac{\phi}{2}\right)$$
$$= \tan^2\left(45° - \frac{25°}{2}\right)$$
$$= 0.406 \quad (0.41)$$

The answer is (B).

2. The passive earth pressure coefficient is

$$K_P = \tan^2\left(45° + \frac{\phi}{2}\right) = \tan^2\left(45° + \frac{25°}{2}\right)$$
$$= 2.46 \quad (2.5)$$

The answer is (C).

3. The active lateral earth pressure coefficient is

$$K_A = \tan^2\left(45° - \frac{\phi}{2}\right)$$
$$= \tan^2\left(45° - \frac{32°}{2}\right)$$
$$= 0.307$$

Find the resultant active force.

$$P_A = \tfrac{1}{2}K_A H^2 \gamma = \left(\tfrac{1}{2}\right)(0.307)(2 \text{ m})^2\left(25 \frac{\text{kN}}{\text{m}^3}\right)$$
$$= 15.4 \text{ kN/m} \quad (15 \text{ kN/m})$$

The answer is (B).

4. The relationships between forces on a free-body diagram of the soil are shown.

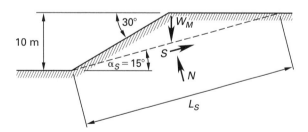

Calculate the length of the assumed planar slip surface.

$$L_S = \frac{h}{\sin\alpha_S} = \frac{10 \text{ m}}{\sin 15°} = 38.64 \text{ m}$$

The height of the triangle wedge, h_t, is

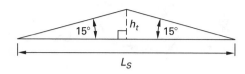

$$h_t = \frac{L_S}{2}\tan\alpha$$
$$= \left(\frac{38.64 \text{ m}}{2}\right)\tan 15°$$
$$= 5.176 \text{ m}$$

Calculate the weight of the soil above the failure plane.

$$W_M = (\text{area of soil wedge})\gamma = \tfrac{1}{2}h_t L_S \gamma$$
$$= \tfrac{1}{2}(5.176 \text{ m})(38.64 \text{ m})\left(14 \frac{\text{kN}}{\text{m}^3}\right)$$
$$= 1400 \text{ kN/m}$$

The available shearing resistance along the failure plane is

$$T_{FF} = cL_S + W_M \cos\alpha_S \tan\phi$$
$$= (15 \text{ kPa})(38.64 \text{ m}) + \left(1400 \frac{\text{kN}}{\text{m}}\right)\cos 15° \tan 15°$$
$$= 942 \text{ kN/m} \quad (940 \text{ kN/m})$$

The answer is (D).

5. The mobilized shear force along the failure plane is

$$T_{\text{MOB}} = W_M \sin\alpha_S$$
$$= \left(2000 \frac{\text{kN}}{\text{m}}\right)\sin 25°$$
$$= 845.2 \text{ kN/m} \quad (850 \text{ kN/m})$$

The answer is (B).

6. The factor of safety against slope instability is

$$FS = \frac{T_{FF}}{T_{\text{MOB}}}$$
$$= \frac{1350 \dfrac{\text{kN}}{\text{m}}}{465 \dfrac{\text{kN}}{\text{m}}}$$
$$= 2.9$$

The answer is (C).

Geotechnical Engineering

7. The coefficient of active lateral earth pressure is

$$K_A = \tan^2\left(45° - \frac{\phi}{2}\right)$$

$$= \tan^2\left(45° - \frac{35°}{2}\right)$$

$$= 0.27$$

The total active force is

$$P_A = \tfrac{1}{2}K_A H^2 \gamma$$

$$= \left(\tfrac{1}{2}\right)(0.27)(3 \text{ m})^2\left(1834 \ \frac{\text{kg}}{\text{m}^3}\right)\left(9.81 \ \frac{\text{m}}{\text{s}^2}\right)$$

$$= 21\,940 \text{ N/m} \quad (22 \text{ kN/m})$$

The answer is (B).

Geotechnical
Engineering

20 Excavations

PRACTICE PROBLEMS

1. An anchored bulkhead is typically supported by an embedded base with a tieback near the top. Sheeting can be used to span the distance between the embedded piles. The LEAST commonly observed failure with this type of construction is typically

- (A) soil heave of the base layer
- (B) failure of the sheeting
- (C) pull-out of the anchorage
- (D) embedment at the base such as toe kick-out

2. For the purposes of calculating the maximum stress in a braced cut, what are the primary factors in determining the shape of the pressure distribution?

- (A) lateral earth pressure coefficient, specific weight, and cut depth
- (B) specific weight, cut depth, and cohesion
- (C) cut depth, soil shear strength, and specific weight
- (D) angle of internal friction, lateral earth pressure coefficient, and cohesion

3. A 30 ft deep, 40 ft square excavation in sand is being designed. The sand will be dewatered before excavation. The angle of internal friction is $40°$, and the specific weight is 121 lbf/ft^3. Bracing consists of horizontal shoring supported by 8 in soldier piles separated horizontally by 8 ft. Assuming a simple uniform pressure distribution behind the shoring, what is most nearly the maximum pressure on the piles?

- (A) 500 lbf/ft^2
- (B) 1100 lbf/ft^2
- (C) 2400 lbf/ft^2
- (D) 3600 lbf/ft^2

SOLUTIONS

1. For anchored bulkheads, the sheeting material spanning between the piles is typically the most reliable portion of the system. The highest failure occurrence is usually in the base heave of soils followed by anchorage pull-out. Failure of the sheeting is rare.

The answer is (B).

2. While the value of the maximum pressure in a braced cut depends on the lateral earth pressure coefficient, specific weight, and cut depth, the shape of the pressure distribution depends on the stability number, $\gamma H/c$.

The answer is (B).

3. The maximum pressure is

$$
\begin{aligned}
p_{\max} &= 0.65 K_A \gamma H \\
&= 0.65 \gamma H \tan^2\left(45° - \frac{\theta}{2}\right) \\
&= (0.65)\left(121\ \frac{lbf}{ft^3}\right)(30\ ft)\tan^2\left(45° - \frac{40°}{2}\right) \\
&= 513\ lbf/ft^2 \quad (500\ lbf/ft^2)
\end{aligned}
$$

The answer is (A).

Geotechnical Engineering

21 Systems of Forces and Moments

PRACTICE PROBLEMS

1. In the system shown, force F, line P, and line Q are coplanar.

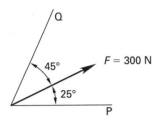

Resolve the 300 N force into two components, one along line P and the other along line Q.

(A) $F_Q = 272$ N; $F_P = 126$ N

(B) $F_Q = 232$ N; $F_P = 186$ N

(C) $F_Q = 135$ N; $F_P = 226$ N

(D) $F_Q = 212$ N; $F_P = 226$ N

2. Which type of load is NOT resisted by a pinned joint?

(A) moment

(B) shear

(C) axial

(D) distributed

3. The loading shown requires a clockwise resisting moment of 20 N·m at the support.

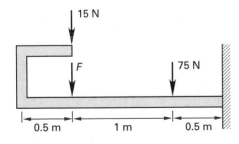

What is most nearly the value of force F?

(A) 25 N (up)

(B) 27 N (up)

(C) 38 N (down)

(D) 43 N (down)

4. A bent beam is acted upon by a moment and several concentrated forces, as shown.

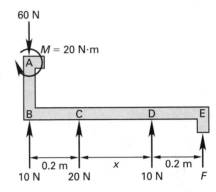

Approximate the unknown force, F, and distance, x, that will maintain equilibrium on the beam.

(A) $F = 5$ N; $x = 0.8$ m

(B) $F = 10$ N; $x = 0.6$ m

(C) $F = 20$ N; $x = 0.2$ m

(D) $F = 20$ N; $x = 0.4$ m

5. For the member shown, the 700 N·m moment is applied at point B.

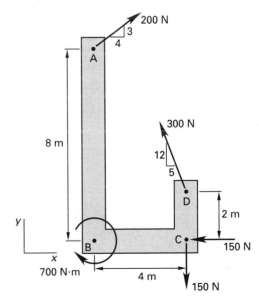

If the member is pinned so that it rotates around point B, what counteracting moment must be applied at point A to keep the member in equilibrium?

(A) 650 N·m

(B) 890 N·m

(C) 1150 N·m

(D) 1240 N·m

6. A force is defined by the vector $\mathbf{A} = 3.5\mathbf{i} - 1.5\mathbf{j} + 2.0\mathbf{k}$. What is most nearly the angle between the force and the positive y-axis?

(A) 20°

(B) 66°

(C) 70°

(D) 110°

7. In the structure shown, the beam is pinned at point B. Point E is a roller support. The beam is loaded with a distributed load from point A to point B of 400 N/m, a 500 N·m couple at point C, and a vertical 900 N force at point D.

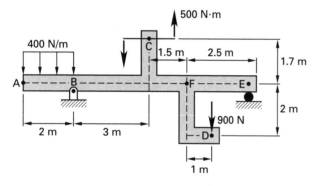

If the distributed load and the vertical load are removed and replaced with a vertically upward force of 1700 N at point F, what moment at point F would be necessary to keep the reaction at point E the same?

(A) −9000 N·m (counterclockwise)

(B) −6500 N·m (counterclockwise)

(C) 3500 N·m (clockwise)

(D) 12 000 N·m (clockwise)

8. Where can a couple be moved on a rigid body to have an equivalent effect?

(A) along the line of action

(B) in a parallel plane

(C) along the perpendicular bisector joining the two original forces

(D) anywhere on the rigid body

9. The overhanging beam shown is supported by a roller and a pinned support. The moment is removed and replaced by a couple consisting of forces applied at points A and C.

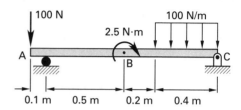

What is the magnitude of the forces that constitute the couple?

(A) 2.1 N

(B) 4.2 N

(C) 6.3 N

(D) 8.3 N

10. A disk-shaped body with a 4 cm radius has a 320 N force acting through the center at an unknown angle θ, and two 40 N loads acting as a couple, as shown. All of these forces are removed and replaced by a single 320 N force at point B, parallel to the original 320 N force.

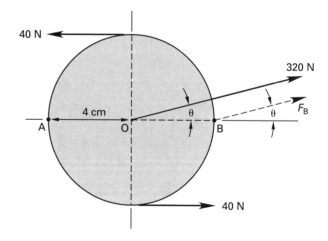

Most nearly, what is the angle θ?

(A) 0°

(B) 7.6°

(C) 15°

(D) 29°

11. The overhanging beam shown is supported by a roller and a pinned support. The moment is removed and replaced by a couple consisting of forces applied at points A and C.

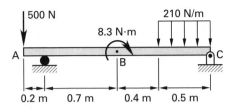

What is most nearly the magnitude of the couple that exactly replaces the moment that is removed?

(A) 0.080 N·m

(B) 0.16 N·m

(C) 8.3 N·m

(D) 15 N·m

12. Which structure is statically determinate and stable with the loadings shown?

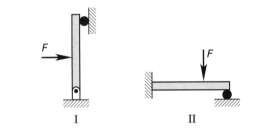

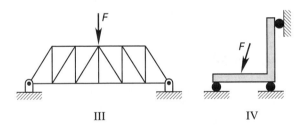

(A) I only

(B) I and III

(C) I and IV

(D) II and III

13. A signal arm carries two traffic signals and a sign, as shown. The signals and sign are rigidly attached to the arm. Each traffic signal has a frontal area of 0.2 m^2 and weighs 210 N. The sign weighs 60 Pa. The design wind pressure is 575 Pa. The maximum moment that the connection between the arm and pole can withstand due to wind is 6000 N·m, and the maximum permitted moment due to the loads is 4000 N·m.

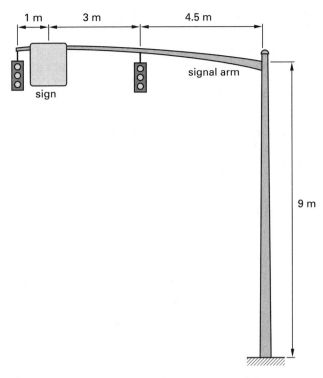

As limited by moment on the connection, what is most nearly the maximum area of the sign?

(A) 1.0 m^2

(B) 1.2 m^2

(C) 2.8 m^2

(D) 5.6 m^2

14. A hinged arch is composed of two pin-connected curved members supported on two pinned supports, as shown. Both members are rigid. A horizontal force of 1000 N is applied to pin B, as shown. All coordinates are in meters.

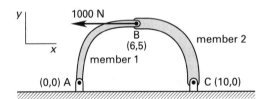

Most nearly, what are the reactions and moments at joint A?

(A) $F_x = 500$ N; $F_y = 600$ N; $M_A = 5000$ N·m

(B) $F_x = 600$ N; $F_y = 500$ N; $M_A = 0$ N·m

(C) $F_x = 680$ N; $F_y = 400$ N; $M_A = 5000$ N·m

(D) $F_x = 616$ N; $F_y = 480$ N; $M_A = 0$ N·m

Statics

SOLUTIONS

1. One of the characteristics of the components of a force is that they combine as vectors into the total force. Draw the vector addition triangle and determine all of the angles and sides.

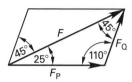

Use the law of sines to calculate the components.

$$\frac{F}{\sin 110°} = \frac{F_Q}{\sin 25°} = \frac{F_P}{\sin 45°}$$

$$\frac{300 \text{ N}}{\sin 110°} = \frac{F_Q}{\sin 25°} = \frac{F_P}{\sin 45°}$$

$$F_Q = (300 \text{ N})\frac{\sin 25°}{\sin 110°} = 134.9 \text{ N} \quad (135 \text{ N})$$

$$F_P = (300 \text{ N})\frac{\sin 45°}{\sin 110°} = 225.7 \text{ N} \quad (226 \text{ N})$$

The answer is (C).

2. A pinned support will resist forces but not moments.

The answer is (A).

3. Clockwise moments are positive. The sum of the moments around the support is

$$\sum M = 0$$
$$= 20 \text{ N·m} - (75 \text{ N})(0.5 \text{ m}) - F(1.5 \text{ m})$$
$$\quad - (15 \text{ N})(1.5 \text{ m})$$
$$F = -26.7 \text{ N} \quad (27 \text{ N}) \quad [\text{up}]$$

The answer is (B).

4. The sum of the forces in the y-direction is

$$\sum F_y = 0$$
$$= -60 \text{ N} + 10 \text{ N} + 20 \text{ N} + 10 \text{ N} + F$$
$$F = 20 \text{ N}$$

Clockwise moments are positive. The sum of the moments around point A is

$$\sum M_A = 0$$
$$= 20 \text{ N·m} - (20 \text{ N})(0.2 \text{ m})$$
$$\quad - (10 \text{ N})(0.2 \text{ m} + x) - (20 \text{ N})(0.4 \text{ m} + x)$$

$$4 + 2 + 10x + 8 + 20x = 20$$
$$30x = 6$$
$$x = 0.2 \text{ m}$$

The answer is (C).

5. Let clockwise moments be positive. Take moments about point B.

$$\sum M_B = 700 \text{ N·m} + (150 \text{ N})(4 \text{ m})$$
$$\quad - (300 \text{ N})\left(\frac{5}{13}\right)(2 \text{ m})$$
$$\quad - (300 \text{ N})\left(\frac{12}{13}\right)(4 \text{ m})$$
$$\quad + (200 \text{ N})\left(\frac{4}{5}\right)(8 \text{ m})$$
$$= 1242 \text{ N·m} \quad (1240 \text{ N·m})$$

The application point of the moment is irrelevant.

The answer is (D).

6. The magnitude of the force is

$$F = \sqrt{F_x^2 + F_y^2 + F_z^2}$$
$$= \sqrt{(3.5)^2 + (-1.5)^2 + (2.0)^2}$$
$$= 4.3$$
$$\theta = \arccos\frac{F_y}{F} = \arccos\frac{-1.5}{4.3}$$
$$= 110.4° \quad (110°)$$

The answer is (D).

7. The reaction at point E is unknown, but it is irrelevant. Since the reaction is to be unchanged, it is necessary only to calculate the change in the loading.

Assume clockwise moments are positive. Take moments about point B for the forces that are removed and added.

$$\sum M_B = -\left(400 \frac{\text{N}}{\text{m}}\right)(2 \text{ m})\left(\frac{2 \text{ m}}{2}\right)$$
$$\quad + (900 \text{ N})(1 \text{ m} + 1.5 \text{ m} + 3 \text{ m})$$
$$\quad - (-1700 \text{ N})(1.5 \text{ m} + 3 \text{ m})$$
$$= 11\,800 \text{ N·m} \quad (12\,000 \text{ N·m}) \quad [\text{clockwise}]$$

This is the moment that is applied by the forces that are removed, reduced by the moment of the new force that is applied. A 12 000 N·m clockwise moment must be applied to counteract the change. The location of the new moment is not relevant.

The answer is (D).

8. Since a couple is composed of two equal but opposite forces, the x- and y-components will always cancel, no matter what the orientation. Only the moment produced by the couple remains.

The answer is (D).

9. The distance between the forces is

$$d = 0.4 \text{ m} + 0.2 \text{ m} + 0.5 \text{ m} + 0.1 \text{ m}$$
$$= 1.2 \text{ m}$$

The forces are

$$F = \frac{M}{d} = \frac{2.5 \text{ N·m}}{1.2 \text{ m}}$$
$$= 2.08 \text{ N} \quad (2.1 \text{ N})$$

The answer is (A).

10. Assume clockwise moments are positive. Take moments about the center for the original forces. The 320 N force has no moment arm, so it does not contribute to the moment. The couple is

$$M = Fd = -(40 \text{ N})(8 \text{ cm})$$
$$= -320 \text{ N·cm}$$

The replacement force must produce a moment of -320 N·cm. The horizontal component of the replacement force acts through the center; only the vertical component of the force contributes to the moment.

$$M = -320 \text{ N·cm} = Fr = -(320 \text{ N})(\sin\theta)(4 \text{ cm})$$
$$\theta = 14.48° \quad (15°)$$

The answer is (C).

11. A couple is a moment. When a moment of 8.3 N·m is removed, it must be replaced by the same moment.

The answer is (C).

12. Structure I is simply supported and determinate. Structure II is a propped cantilever beam, always indeterminate by one degree. Structure III is a truss that is pinned at both ends, also indeterminate by one degree. Structure IV is a beam with three rollers, two in the vertical direction and one in the horizontal direction. It is determinate, but not stable.

The answer is (A).

13. The length of the signal arm is

$$1 \text{ m} + 3 \text{ m} + 4.5 \text{ m} = 8.5 \text{ m}$$

Set the moment on the arm due to the wind equal to the maximum allowed.

$$\sum M_{\text{wind}} = (0.2 \text{ m}^2)(575 \text{ Pa})(8.5 \text{ m} + 4.5 \text{ m})$$
$$+ A_{\text{sign}}(575 \text{ Pa})(7.5 \text{ m}) = 6000 \text{ N·m}$$
$$A_{\text{sign}} = 1.04 \text{ m}^2$$

Set the moment on the arm due to vertical loading equal to the maximum allowed.

$$\sum M_{\text{loads}} = (210 \text{ N})(8.5 \text{ m} + 4.5 \text{ m})$$
$$+ (60 \text{ Pa})A_{\text{sign}}(7.5 \text{ m}) = 4000 \text{ N·m}$$
$$A_{\text{sign}} = 2.82 \text{ m}^2$$

The maximum area of the sign is the smaller of these two values, 1.04 m^2 (1.0 m^2).

The answer is (A).

14. Sum moments about point C. The x-component of the reaction at point A has a zero moment arm.

$$\sum M_{\text{C}} = 10F_y - (1000 \text{ N})(5) = 0$$
$$F_y = 500 \text{ N}$$

Point B is a pin, which transmits no moment. Sum moments to the left of point B.

$$\sum M_{\text{B}} = F_y(6) - F_x(5) = 0$$
$$= (500 \text{ N})(6) - F_x(5) = 0$$
$$F_x = 600 \text{ N}$$

Point A is a pin, which transmits no moment.

The answer is (B).

Statics

22 Trusses

PRACTICE PROBLEMS

1. Determine the approximate force in member BC for the truss shown.

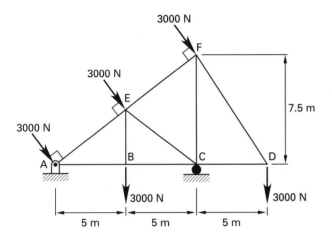

(A) 0 N

(B) 1000 N

(C) 1500 N

(D) 2500 N

2. Find the approximate force in member DE for the truss shown.

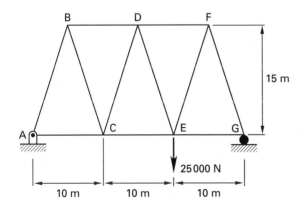

(A) 0 N

(B) 6300 N

(C) 8800 N

(D) 10 000 N

3. For the truss shown, what are most nearly the forces in members AC and BD?

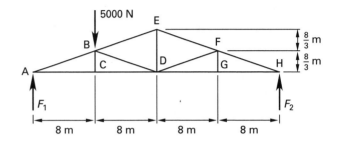

(A) $AC = 11\,000$ N; $BD = -7900$ N

(B) $AC = 0$ N; $BD = -2000$ N

(C) $AC = 1100$ N; $BD = 2500$ N

(D) $AC = 0$ N; $BD = -7900$ N

4. The braced frame shown is constructed with pin-connected members and supports. All applied forces are horizontal.

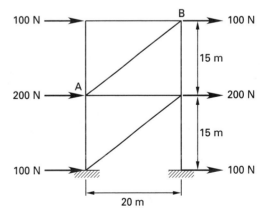

Most nearly, what is the force in the diagonal member BA?

(A) 0 N

(B) 160 N

(C) 200 N

(D) 250 N

5. For the cantilever truss shown, what is most nearly the force in member AF?

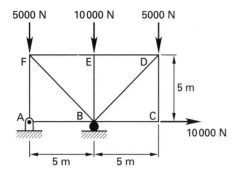

(A) 0 N

(B) 5000 N

(C) 10 000 N

(D) 15 000 N

SOLUTIONS

1. Distance AF is

$$AF = \sqrt{(5 \text{ m} + 5 \text{ m})^2 + (7.5 \text{ m})^2} = 12.5 \text{ m}$$

Distance AE is

$$AE = \tfrac{1}{2}AF = \left(\tfrac{1}{2}\right)(12.5 \text{ m}) = 6.25 \text{ m}$$

The sum of the moments around point A is

$$\sum M_A = 0$$
$$= (3000 \text{ N})(6.25 \text{ m}) + (3000 \text{ N})(12.5 \text{ m})$$
$$+ (3000 \text{ N})(5 \text{ m}) - F_{C_y}(10 \text{ m})$$
$$+ (3000 \text{ N})(15 \text{ m})$$
$$F_{C_y} = 11\,625 \text{ N} \quad \text{[upward]}$$

From trigonometry, the applied forces are inclined from the horizontal at $\arctan(7.5 \text{ m}/(5 \text{ m} + 5 \text{ m})) = 36.87°$.

$$\sum F_y = 0$$
$$= F_{A_y} - (3)(3000 \text{ N})\cos 36.87°$$
$$- (2)(3000 \text{ N}) + 11\,625 \text{ N}$$
$$F_{A_y} = 1575 \text{ N} \quad \text{[upward]}$$
$$\sum F_x = 0$$
$$= F_{A_x} + (3)(3000 \text{ N})\sin 36.87°$$
$$F_{A_x} = -5400 \text{ N} \quad \text{[to the left]}$$

Use the method of sections.

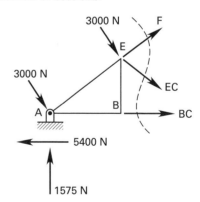

Take moments about point E. The vertical downward force at point B passes through point E, and it does not generate a moment.

$$\sum M_E = 0$$
$$= (5400 \text{ N})(3.75 \text{ m}) + (1575 \text{ N})(5 \text{ m})$$
$$- (3000 \text{ N})(6.25 \text{ m}) - BC(3.75 \text{ m})$$
$$BC = 2500 \text{ N}$$

The answer is (D).

2. Take moments about point G.

$$\sum M_{\mathrm{G}} = 0$$
$$= (-25\,000 \text{ N})(10 \text{ m}) + F_{\mathrm{A}_y}(30 \text{ m})$$
$$F_{\mathrm{A}_y} = 8333 \text{ N} \quad \text{[upward]}$$

Use the method of sections.

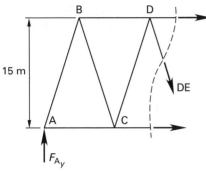

$$\sum F_y = 0$$
$$= 8333 \text{ N} - \mathrm{DE}_y$$
$$\mathrm{DE}_y = 8333 \text{ N}$$
$$\mathrm{DE}_x = (8333 \text{ N})\left(\frac{5 \text{ m}}{15 \text{ m}}\right)$$
$$= 2778 \text{ N}$$
$$\mathrm{DE} = \sqrt{(8333 \text{ N})^2 + (2778 \text{ N})^2}$$
$$= 8784 \text{ N} \quad (8800 \text{ N}) \quad \text{[tension]}$$

The answer is (C).

3. Take moments about point H.

$$\sum M_{\mathrm{H}} = 0$$
$$= (5000 \text{ N})(24 \text{ m}) - F_1(32 \text{ m})$$
$$F_1 = 3750 \text{ N}$$

The angle made by the inclined members with the horizontal is

$$\arctan \frac{\frac{8}{3} \text{ m}}{8 \text{ m}} = 18.435°$$

(Alternatively, the force components could be determined from geometry.)

Use the method of joints.

For pin A,

$$\sum F_y = 0$$
$$= F_1 + \mathrm{AB} \sin 18.435°$$
$$= 3750 \text{ N} + \mathrm{AB} \sin 18.435°$$
$$\mathrm{AB} = -11\,859 \text{ N} \quad \text{[compression]}$$
$$\sum F_x = 0$$
$$= (-11\,859 \text{ N})\cos 18.435° + \mathrm{AC}$$
$$\mathrm{AC} = 11\,250 \text{ N} \quad (11\,000 \text{ N}) \quad \text{[tension]}$$

For pin C,

$$\sum F_y = 0$$
$$\mathrm{BC} = 0 \quad \text{[zero-force member]}$$

For pin B,

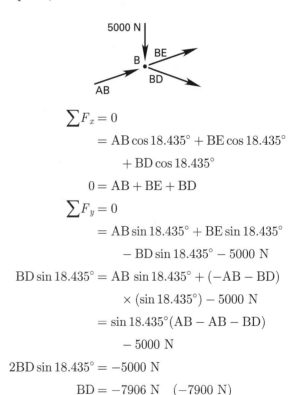

$$\sum F_x = 0$$
$$= \mathrm{AB} \cos 18.435° + \mathrm{BE} \cos 18.435°$$
$$\quad + \mathrm{BD} \cos 18.435°$$
$$0 = \mathrm{AB} + \mathrm{BE} + \mathrm{BD}$$
$$\sum F_y = 0$$
$$= \mathrm{AB} \sin 18.435° + \mathrm{BE} \sin 18.435°$$
$$\quad - \mathrm{BD} \sin 18.435° - 5000 \text{ N}$$
$$\mathrm{BD} \sin 18.435° = \mathrm{AB} \sin 18.435° + (-\mathrm{AB} - \mathrm{BD})$$
$$\quad \times (\sin 18.435°) - 5000 \text{ N}$$
$$= \sin 18.435°(\mathrm{AB} - \mathrm{AB} - \mathrm{BD})$$
$$\quad - 5000 \text{ N}$$
$$2\mathrm{BD} \sin 18.435° = -5000 \text{ N}$$
$$\mathrm{BD} = -7906 \text{ N} \quad (-7900 \text{ N})$$
$$\begin{bmatrix} \text{compression in opposite} \\ \text{direction shown} \end{bmatrix}$$

The answer is (A).

Statics

4. Determine the length of member BA by recognizing this configuration to be a 3-4-5 triangle.

$$L_{BA} = 25 \text{ m}$$

Use the method of sections. Cut the frame horizontally through member BA.

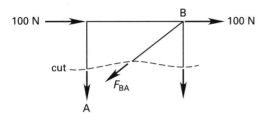

By inspection, the horizontal component of F_{BA} balances the two applied horizontal loads.

$$BA_x = 100 \text{ N} + 100 \text{ N} = 200 \text{ N}$$

By similar triangles,

$$BA = \left(\tfrac{5}{4}\right)(200 \text{ N}) = 250 \text{ N}$$

The answer is (D).

5. Take moments about point B.

$$\sum M_B = 0$$
$$= -(5000 \text{ N})(5 \text{ m}) + (5000 \text{ N})(5 \text{ m})$$
$$+ R_{A_y}(5 \text{ m})$$
$$R_{A_y} = 0 \text{ N}$$

By inspection, the force in member AF is equal to the vertical component of the reaction at point A.

The answer is (A).

Statics

23 Pulleys, Cables, and Friction

PRACTICE PROBLEMS

1. A 2 kg block rests on a 34° incline.

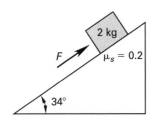

If the coefficient of static friction is 0.2, approximately how much additional force, F, must be applied to keep the block from sliding down the incline?

(A) 7.7 N

(B) 8.8 N

(C) 9.1 N

(D) 14 N

2. A box has uniform density and a total weight of 600 N. It is suspended by three equal-length cables, AE, BE, and CE, as shown. Point E is 0.5 m directly above the center of the box's top surface.

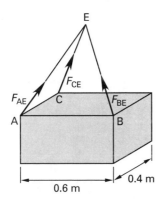

Most nearly, what is the tension in cable CE?

(A) 130 N

(B) 200 N

(C) 370 N

(D) 400 N

3. A rope is wrapped over a 6 cm diameter pipe to support a bucket of tools being lowered. The coefficient of static friction between the rope and the pipe is 0.20. The combined mass of the bucket and tools is 100 kg.

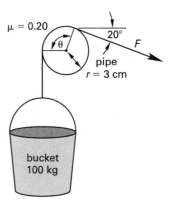

What is most nearly the range of force that can be applied to the free end of the rope such that the bucket remains stationary?

(A) 560 N to 1360 N

(B) 670 N to 1440 N

(C) 720 N to 1360 N

(D) 720 N to 1510 N

4. The two cables shown carry a 100 N vertical load.

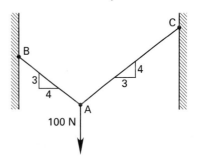

Most nearly, what is the tension in cable AB?

(A) 40 N

(B) 50 N

(C) 60 N

(D) 80 N

5. Most nearly, what is the tension, T, that must be applied to pulley A to lift the 1200 N weight?

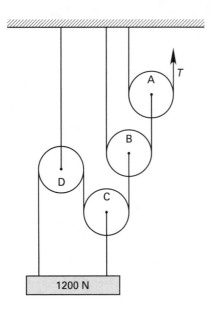

(A) 75 N

(B) 100 N

(C) 300 N

(D) 400 N

SOLUTIONS

1. Choose coordinate axes parallel and perpendicular to the incline.

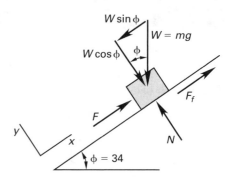

The sum of the forces is

$$\sum F_x = 0 = F + F_f - W \sin \phi$$

$$F = W \sin \phi - F_f$$

$$= mg \sin \phi - \mu_s N$$

$$= mg \sin \phi - \mu_s mg \cos \phi$$

$$= mg(\sin \phi - \mu \cos \phi)$$

$$= (2 \text{ kg}) \left(9.81 \ \tfrac{\text{m}}{\text{s}^2}\right) (\sin 34° - 0.2 \cos 34°)$$

$$= 7.7 \text{ N}$$

The answer is (A).

2. The length of the diagonal is

$$\text{BC} = \sqrt{(0.4 \text{ m})^2 + (0.6 \text{ m})^2} = 0.721 \text{ m}$$

The cable length is

$$\text{BE} = \sqrt{\left(\frac{0.721 \text{ m}}{2}\right)^2 + (0.5 \text{ m})^2} = 0.616 \text{ m}$$

There is nothing to balance the component force in the direction from point A to the opposite corner, so the force in cable AE is zero. Cables BE and CE each carry half of the box weight. The vertical component of force in each cable is 300 N.

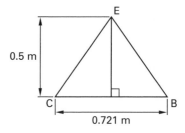

By similar triangles, the tensile force in each cable is

$$T = \frac{(300 \text{ N})(0.616 \text{ m})}{0.5 \text{ m}} = 370 \text{ N}$$

The answer is (C).

3. The angle of wrap is

$$\theta = (90° + 20°)\left(\frac{2\pi \text{ rad}}{360°}\right) = 1.92 \text{ rad}$$

(This must be expressed in radians.)

The tensile force in the rope due to the bucket's mass is

$$F = mg = (100 \text{ kg})\left(9.81 \frac{\text{m}}{\text{s}^2}\right)$$
$$= 981 \text{ N}$$

The free end of the rope can either be on the tight or loose side. These two options define the range of force that will keep the bucket stationary.

The ratio of tight-side to loose-side forces is

$$\frac{F_1}{F_2} = e^{\mu\theta} = e^{(0.20)(1.92 \text{ rad})}$$
$$= 1.468$$

Multiply and divide the bucket-end tension by this ratio.

$$\text{minimum tension:} \quad \frac{981 \text{ N}}{1.468} = 668 \text{ N} \quad (670 \text{ N})$$
$$\text{maximum tension:} \quad (1.468)(981 \text{ N}) = 1440 \text{ N}$$

The answer is (B).

4. Recognize that the orientations of both cables are defined by 3-4-5 triangles.

The equilibrium condition for horizontal forces at point A is

$$F_x = T_{AC} - T_{AB} = 0$$
$$\tfrac{3}{5}T_{AC} - \tfrac{4}{5}T_{AB} = 0$$
$$T_{AC} = \tfrac{4}{3}T_{AB}$$

The equilibrium condition for vertical forces at point A is

$$F_y = T_{AB} + T_{AC} - 100 \text{ N} = 0$$
$$\tfrac{3}{5}T_{AB} + \tfrac{4}{5}T_{AC} - 100 \text{ N} = 0$$

Substitute $(4/3)\,T_{AB}$ for T_{AC}.

$$\tfrac{3}{5}T_{AB} + \left(\tfrac{4}{5}\right)\left(\tfrac{4}{3}\right)T_{AB} = 100 \text{ N}$$
$$\left(\tfrac{3}{5} + \tfrac{16}{15}\right)T_{AB} = 100 \text{ N}$$
$$T_{AB} = 60 \text{ N}$$

The answer is (C).

5. The free bodies of the system are shown.

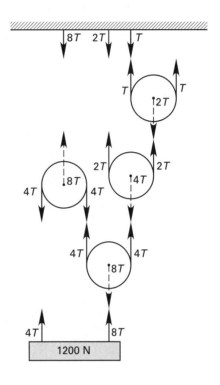

$$\sum F_y = 0$$
$$= -1200 \text{ N} + 4T + 8T$$
$$12T = 1200 \text{ N}$$
$$T = 100 \text{ N}$$

The answer is (B).

24 Centroids and Moments of Inertia

PRACTICE PROBLEMS

1. Refer to the complex shape shown.

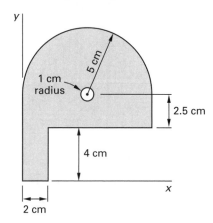

What is most nearly the moment of inertia about the x-axis?

(A) 1500 cm^4

(B) 3400 cm^4

(C) 4600 cm^4

(D) 5200 cm^4

2. Refer to the composite plane areas shown.

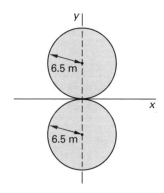

What is the approximate polar moment of inertia about the composite centroid?

(A) 2400 m^4

(B) 5500 m^4

(C) $12\,000 \text{ m}^4$

(D) $17\,000 \text{ m}^4$

3. Most nearly, what is the area moment of inertia about the x-axis for the area shown?

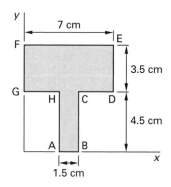

(A) 89 cm^4

(B) 170 cm^4

(C) 510 cm^4

(D) 1000 cm^4

4. Refer to the cross section of the angle shown.

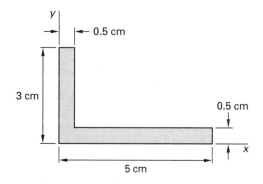

Most nearly, what is the x-coordinate of the centroid of the perimeter line?

(A) 1.56 cm

(B) 1.66 cm

(C) 1.75 cm

(D) 1.80 cm

5. The centroidal moment of inertia about the x-axis for the area shown is 142.41 cm^4.

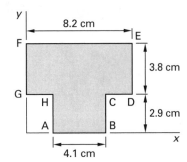

Most nearly, what is the centroidal polar moment of inertia?

(A) 79 cm^4

(B) 110 cm^4

(C) 330 cm^4

(D) 450 cm^4

6. Refer to the complex shape shown.

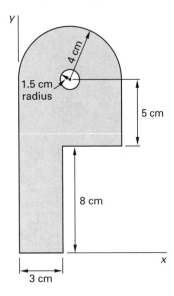

What is most nearly the y-coordinate of the centroid?

(A) 5.5 cm

(B) 7.3 cm

(C) 9.2 cm

(D) 11 cm

7. Most nearly, what are the x- and y-coordinates of the centroid of the perimeter line for the area shown?

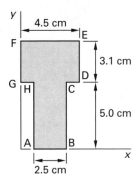

(A) 1.0 cm; 3.8 cm

(B) 1.0 cm; 4.0 cm

(C) 2.3 cm; 4.4 cm

(D) 2.3 cm; 4.8 cm

8. Refer to the composite plane areas shown. The polar moment of inertia is $3\pi r^4$.

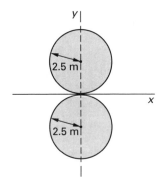

What is most nearly the polar radius of gyration?

(A) 2.5 m

(B) 2.7 m

(C) 2.9 m

(D) 3.1 m

9. The y-coordinate of the centroid of the area shown is 5.8125 cm.

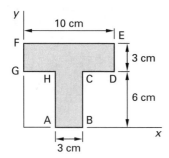

Most nearly, what is the centroidal moment of inertia with respect to the x-axis?

(A) 82 cm^4

(B) 100 cm^4

(C) 220 cm^4

(D) 270 cm^4

10. Refer to the complex shape shown.

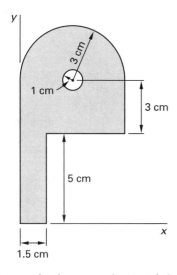

What is most nearly the x-coordinate of the centroid?

(A) 2.4 cm

(B) 2.5 cm

(C) 2.8 cm

(D) 3.2 cm

11. The centroidal moment of inertia of area A_2 with respect to the composite centroidal x-axis is 73.94 cm^4. The y-coordinate of the composite centroid is 4.79 cm. The moment of inertia of area A_3 with respect to the composite centroidal x-axis is 32.47 cm^4.

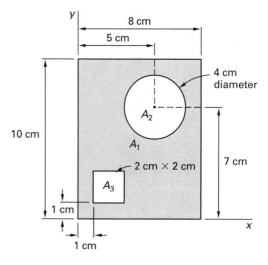

Most nearly, what is the moment of inertia of the entire composite area with respect to its centroidal x-axis?

(A) 350 cm^4

(B) 460 cm^4

(C) 480 cm^4

(D) 560 cm^4

12. Most nearly, what are the x- and y-coordinates of the centroid of the area shown?

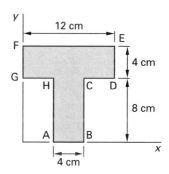

(A) 4.8 cm; 6.8 cm

(B) 6.0 cm; 7.2 cm

(C) 6.0 cm; 7.6 cm

(D) 6.0 cm; 8.0 cm

13. Refer to the complex shape shown. The moment of inertia about the y-axis is 352 cm^4, and the total area of the shape is 36.5 cm^2.

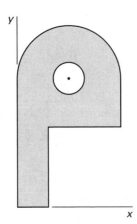

What is most nearly the radius of gyration with respect to the y-axis?

- (A) 1.9 cm
- (B) 3.1 cm
- (C) 3.3 cm
- (D) 4.0 cm

SOLUTIONS

1. Calculate the centroidal moment of inertia, the area, and the distance from that centroid to the x-axis for each region.

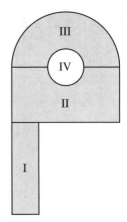

For region I (rectangular),

$$A = (2 \text{ cm})(4 \text{ cm}) = 8 \text{ cm}^2$$

$$I_c = \frac{bh^3}{12} = \frac{(2 \text{ cm})(4 \text{ cm})^3}{12} = 10.667 \text{ cm}^4$$

$$d = \frac{4 \text{ cm}}{2} = 2 \text{ cm}$$

$$d^2 A = (2 \text{ cm})^2 (8 \text{ cm}^2) = 32 \text{ cm}^4$$

For region II (rectangular, including half of region IV),

$$A = (2)(5 \text{ cm})(2.5 \text{ cm}) = 25 \text{ cm}^2$$

$$I_c = \frac{bh^3}{12} = \frac{(8 \text{ cm})(2.5 \text{ cm})^3}{12} = 10.417 \text{ cm}^4$$

$$d = 4 \text{ cm} + \frac{2.5 \text{ cm}}{2} = 5.25 \text{ cm}$$

$$d^2 A = (5.25 \text{ cm})^2 (25 \text{ cm}^2) = 689.0625 \text{ cm}^4$$

For region III (semicircular, including half of region IV),

$$A = \frac{\pi r^2}{2} = \frac{\pi (5 \text{ cm})^2}{2} = \frac{25\pi}{2}$$

$$I_c = 0.1098 r^4 = (0.1098)(5 \text{ cm})^4 = 68.625 \text{ cm}^4$$

$$d = 4 \text{ cm} + 2.5 \text{ cm} + \frac{4r}{3\pi}$$

$$= 6.5 \text{ cm} + \frac{(4)(5 \text{ cm})}{3\pi}$$

$$= 6.5 \text{ cm} + \frac{20}{3\pi} \text{ cm}$$

$$d^2 A = \left(6.5 \text{ cm} + \frac{20}{3\pi} \text{ cm}\right)^2 \left(\frac{25\pi}{2} \text{ cm}^2\right) = 2919.3 \text{ cm}^4$$

For region IV (circular),

$$A = \pi r^2 = \pi (1 \text{ cm})^2 = \pi \text{ cm}^2$$

$$I_c = \frac{\pi r^4}{4} = \frac{\pi (1 \text{ cm})^4}{4}$$

$$= \frac{\pi}{4} \text{ cm}^4$$

$$d = 4 \text{ cm} + 2.5 \text{ cm}$$

$$= 10 \text{ cm}$$

$$d^2 A = (10 \text{ cm})^2 (\pi \text{ cm}^2)$$

$$= 100\pi \text{ cm}^4$$

Use the parallel axis theorem for each shape.

$$I_x = \sum I_{x_c} + \sum d^2 A$$

$$= 10.667 \text{ cm}^4 + 32 \text{ cm}^4$$

$$+ 10.417 \text{ cm}^4 + 689.0625 \text{ cm}^4$$

$$+ 68.625 \text{ cm}^4 + 2919.3 \text{ cm}^4$$

$$- \frac{\pi}{4} \text{ cm}^4 - 100\pi \text{ cm}^4$$

$$= 3415 \text{ cm}^4 \quad (3400 \text{ cm}^4)$$

The answer is (B).

2. For a circle,

$$I_{x_c} = I_{y_c} = \frac{\pi r^4}{4}$$

Use the parallel axis theorem for a composite area.

$$I_{x_c} = (2)\left(\frac{\pi r^4}{4} + (\pi r^2) r^2\right) = \frac{5\pi r^4}{2}$$

$$I_{y_c} = (2)\left(\frac{\pi r^4}{4}\right) = \frac{\pi r^4}{2}$$

$$J_c = I_{x_c} + I_{y_c} = \frac{5\pi r^4}{2} + \frac{\pi r^4}{2}$$

$$= 3\pi r^4 = 3\pi (6.5 \text{ m})^4$$

$$= 16\,824 \text{ m}^4 \quad (17\,000 \text{ m}^4)$$

The answer is (D).

3. Use the formula for the moment of inertia about an edge for a rectangular shape.

For rectangle HCBA,

$$I_{x,1} = \frac{bh^3}{3} = \frac{(1.5 \text{ cm})(4.5 \text{ cm})^3}{3}$$

$$= 45.56 \text{ cm}^4$$

($I_c = bh^3/12$ could have been used, but the parallel axis theorem would also have to be used.)

Use the parallel axis theorem to calculate the moment of inertia of rectangle FEDG. $d = 6.25$ cm is the distance from the centroid of FEDG to the x-axis.

$$I_{x,2} = \frac{bh^3}{12} + A d^2$$

$$= \frac{(7 \text{ cm})(3.5 \text{ cm})^3}{12} + (7 \text{ cm})(3.5 \text{ cm})(6.25 \text{ cm})^2$$

$$= 982 \text{ cm}^4$$

The moment of inertia for the total area is

$$I_x = I_{x,1} + I_{x,2}$$

$$= 45.56 \text{ cm}^4 + 982 \text{ cm}^4$$

$$= 1028 \text{ cm}^4 \quad (1000 \text{ cm}^4)$$

The answer is (D).

4. The length of the perimeter is

$$L = 5 \text{ cm} + 0.5 \text{ cm} + 4.5 \text{ cm}$$

$$+ 2.5 \text{ cm} + 0.5 \text{ cm} + 3 \text{ cm}$$

$$= 16 \text{ cm}$$

The x-coordinate of the centroid of the line is

$$x_c = \frac{\sum x_i L}{L}$$

$$= \left(\frac{1}{16 \text{ cm}}\right)\big((2.5 \text{ cm})(5 \text{ cm}) + (5 \text{ cm})(0.5 \text{ cm})$$

$$+ (2.75 \text{ cm})(5 \text{ cm} - 0.5 \text{ cm})$$

$$+ (0.5 \text{ cm})(3 \text{ cm} - 0.5 \text{ cm})$$

$$+ (0.25 \text{ cm})(0.5 \text{ cm}) + (0 \text{ cm})(3 \text{ cm})\big)$$

$$= 1.80 \text{ cm}$$

The answer is (D).

5. The y-axis passes through the centroid of the area, so the parallel axis theorem is not needed. Since the centroids of the individual rectangles coincide with the centroid of the composite area about the y-axis, I_{y_c} is simply the sum of the moments of inertia of the individual areas.

$$I_{y_c} = \frac{(4.1 \text{ cm})^3 (2.9 \text{ cm})}{12} + \frac{(8.2 \text{ cm})^3 (3.8 \text{ cm})}{12}$$

$$= 191.26 \text{ cm}^4$$

Statics

The centroidal polar moment of inertia is

$$J_c = I_{x_c} + I_{y_c}$$
$$= 141.41 \text{ cm}^4 + 191.26 \text{ cm}^4$$
$$= 332.7 \text{ cm}^4 \quad (330 \text{ cm}^4)$$

The answer is (C).

6. Separate the shape into regions. Calculate the area and locate the centroid for each region.

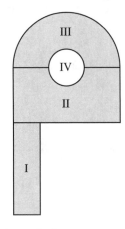

For region I (rectangular),

$$A = (3 \text{ cm})(8 \text{ cm})$$
$$= 24 \text{ cm}^2$$
$$y_c = \frac{8 \text{ cm}}{2}$$
$$= 4 \text{ cm}$$

For region II (rectangular, including half of region IV),

$$A = (2)(4 \text{ cm})(5 \text{ cm})$$
$$= 40 \text{ cm}^2$$
$$y = 8 \text{ cm} + \frac{5 \text{ cm}}{2}$$
$$= 10.5 \text{ cm}$$

For region III (semicircular, including half of region IV),

$$A = \frac{\pi r^2}{2} = \frac{\pi (4 \text{ cm})^2}{2}$$
$$= 8\pi \text{ cm}^2$$
$$y = 8 \text{ cm} + 5 \text{ cm} + \frac{4r}{3\pi}$$
$$= 13 \text{ cm} + \frac{(4)(4 \text{ cm})}{3\pi}$$
$$= 13 \text{ cm} + \frac{16}{3\pi} \text{ cm}$$

For region IV (circular),

$$A = \pi r^2 = \pi (1.5 \text{ cm})^2$$
$$= 2.25\pi \text{ cm}^2$$
$$y = 8 \text{ cm} + 5 \text{ cm} = 13 \text{ cm}$$

$$y_c = \frac{\sum y_{c,n} A_n}{\sum A_n}$$

$$= \frac{\begin{array}{c}(4 \text{ cm})(24 \text{ cm}^2) + (10.5 \text{ cm})(40 \text{ cm}^2) \\ + \left(13 \text{ cm} + \frac{16}{3\pi} \text{ cm}\right)(8\pi \text{ cm}^2) \\ - (13 \text{ cm})(2.25\pi \text{ cm}^2)\end{array}}{24 \text{ cm}^2 + 40 \text{ cm}^2 + 8\pi \text{ cm}^2 - 2.25\pi \text{ cm}^2}$$

$$= \frac{1067 \text{ cm}^3}{96.20 \text{ cm}^2}$$
$$= 11.09 \text{ cm} \quad (11 \text{ cm})$$

The answer is (D).

7. The total length of the perimeter is

$$L = 2.5 \text{ cm} + 5.0 \text{ cm} + 1.0 \text{ cm} + 3.1 \text{ cm}$$
$$+ 4.5 \text{ cm} + 3.1 \text{ cm} + 1.0 \text{ cm} + 5.0 \text{ cm}$$
$$= 25.2 \text{ cm}$$

The x-coordinate of the centroid of the perimeter is

$$x_c = \frac{\sum x_i L_i}{L}$$

$$= \frac{\begin{array}{c}(0 \text{ cm})(3.1 \text{ cm}) + (0.5 \text{ cm})(1.0 \text{ cm}) \\ + (1.0 \text{ cm})(5.0 \text{ cm}) \\ + (2.25 \text{ cm})(4.5 \text{ cm} + 2.5 \text{ cm}) \\ + (3.5 \text{ cm})(5.0 \text{ cm}) \\ + (4 \text{ cm})(1 \text{ cm}) \\ + (4.5 \text{ cm})(3.1 \text{ cm})\end{array}}{25.2 \text{ cm}}$$

$$= 2.25 \text{ cm} \quad (2.3 \text{ cm})$$

The y-coordinate is

$$y_c = \sum y_i L_i$$

$$= \frac{\begin{array}{c}(0 \text{ cm})(2.5 \text{ cm}) + (2)(2.5 \text{ cm})(5.0 \text{ cm}) \\ + (2)(5.0 \text{ cm})(1 \text{ cm}) + (2)(6.55 \text{ cm})(3.1 \text{ cm}) \\ + (8.1 \text{ cm})(4.5 \text{ cm})\end{array}}{25.2 \text{ cm}}$$

$$= 4.447 \text{ cm} \quad (4.4 \text{ cm})$$

The answer is (C).

8. The polar radius of gyration is

$$r_p = \sqrt{\frac{J}{A}} = \sqrt{\frac{3\pi r^4}{2\pi r^2}}$$

$$= r\sqrt{\frac{3}{2}} = (2.5 \text{ m})\sqrt{\frac{3}{2}}$$

$$= 3.1 \text{ m}$$

The answer is (D).

9. Use the parallel axis theorem to find the centroidal moment of inertia of each rectangular area.

$$\text{HCBA} = \text{area 1} = (3 \text{ cm})(6 \text{ cm}) = 18 \text{ cm}^2$$
$$\text{FEDG} = \text{area 2} = (10 \text{ cm})(3 \text{ cm}) = 30 \text{ cm}^2$$

$$I_{x_c} = (I_{c,1} + A_1 d_1^2) + (I_{c,2} + A_2 d_2^2)$$

$$= \left(\begin{array}{c} \dfrac{(3 \text{ cm})(6 \text{ cm})^3}{12} + (18 \text{ cm}^2) \\ \times (5.8125 \text{ cm} - 3.0 \text{ cm})^2 \end{array} \right)$$

$$= \left(\begin{array}{c} \dfrac{(10 \text{ cm})(3 \text{ cm})^3}{12} + (30 \text{ cm}^2) \\ \times (7.5 \text{ cm} - 5.8125 \text{ cm})^2 \end{array} \right)$$

$$= 269.5 \text{ cm}^4 \quad (270 \text{ cm}^4)$$

The answer is (D).

10. Separate the shape into regions. Calculate the area and locate the centroid for each region.

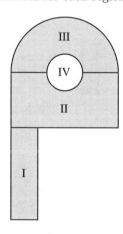

For region I (rectangular),

$$A = (1.5 \text{ cm})(5 \text{ cm})$$

$$= 7.5 \text{ cm}^2$$

$$x_c = \frac{1.5 \text{ cm}}{2}$$

$$= 0.75 \text{ cm}$$

For region II (rectangular, including half of region IV),

$$A = (2)(3 \text{ cm})(3 \text{ cm})$$

$$= 18 \text{ cm}^2$$

$$x_c = \frac{(2)(3 \text{ cm})}{2}$$

$$= 3 \text{ cm}$$

For region III (semicircular, including half of region IV),

$$A = \frac{\pi r^2}{2} = \frac{\pi(3 \text{ cm})^2}{2}$$

$$= \frac{9\pi}{2} \text{ cm}^2$$

$$x_c = 3 \text{ cm} \quad \text{[by inspection]}$$

For region IV (circular),

$$A = \pi r^2 = \pi(1 \text{ cm})^2$$

$$= \pi \text{ cm}^2$$

$$x_c = 3 \text{ cm} \quad \text{[by inspection]}$$

For the total area,

$$x_c = \frac{\sum x_{c,n} A_n}{\sum A_n}$$

$$= \frac{\begin{array}{c}(0.75 \text{ cm})(7.5 \text{ cm}^2) + (3 \text{ cm})(18 \text{ cm}^2) \\ + (3 \text{ cm})\left(\dfrac{9\pi}{2} \text{ cm}^2\right) - (3 \text{ cm})(\pi \text{ cm}^2)\end{array}}{7.5 \text{ cm}^2 + 18 \text{ cm}^2 + \dfrac{9\pi}{2} \text{ cm}^2 - \pi \text{ cm}^2}$$

$$= 2.54 \text{ cm} \quad (2.5 \text{ cm})$$

The answer is (B).

11. The vertical distance between the centroidal location of area A_1 and the composite area's centroid is

$$d_1 = \frac{h}{2} - y_c = \frac{10 \text{ cm}}{2} - 4.79 \text{ cm}$$

The moment of inertia is

$$I_{x_c} = (I_{x_c,1} + A_1 d_1^2) - I_{x'_c,2} - I_{x'_c,3}$$

$$= \left(\begin{array}{c} \dfrac{(8 \text{ cm})(10 \text{ cm})^3}{12} \\ + (8 \text{ cm})(10 \text{ cm})\left(\dfrac{10 \text{ cm}}{2} - 4.79 \text{ cm}\right)^2 \end{array} \right)$$

$$- 73.94 \text{ cm}^4 - 32.47 \text{ cm}^4$$

$$= 563.8 \text{ cm}^4 \quad (560 \text{ cm}^4)$$

The answer is (D).

Statics

12. Divide the area into two rectangles, HCBA and FEDG. Their areas are

$$A_1 = (4 \text{ cm})(8 \text{ cm}) = 32 \text{ cm}^2$$
$$A_2 = (12 \text{ cm})(4 \text{ cm}) = 48 \text{ cm}^2$$

The total area is

$$A = A_1 + A_2 = 32 \text{ cm}^2 + 48 \text{ cm}^2$$
$$= 80 \text{ cm}^2$$

By inspection, the x-coordinates of the centroids of the rectangles are $x_{c,1} = x_{c,2} = 6$ cm. The x-coordinate of the centroid of the total area is

$$x_c = \frac{\sum x_{c,n} A_n}{A}$$
$$= \frac{(6 \text{ cm})(32 \text{ cm}^2) + (6 \text{ cm})(48 \text{ cm}^2)}{80 \text{ cm}^2}$$
$$= 6.0 \text{ cm}$$

By inspection, the y-coordinates of the rectangles are $y_{c,1} = 4$ cm and $y_{c,2} = 10$ cm. The y-coordinate of the centroid of the total area is

$$y_c = \frac{\sum y_{c,n} A_n}{A}$$
$$= \frac{(4 \text{ cm})(32 \text{ cm}^2) + (10 \text{ cm})(48 \text{ cm}^2)}{80 \text{ cm}^2}$$
$$= 7.6 \text{ cm}$$

The answer is (C).

13. The radius of gyration with respect to the y-axis is

$$r_y = \sqrt{\frac{I_y}{A}} = \sqrt{\frac{352 \text{ cm}^4}{36.5 \text{ cm}^2}}$$
$$= 3.1 \text{ cm}$$

The answer is (B).

Statics

25 Indeterminate Statics

PRACTICE PROBLEMS

1. What are the approximate vertical reactions at the ends of the structure shown?

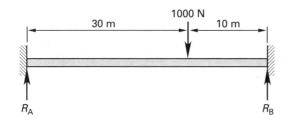

(A) $R_A = 120$ N; $R_B = 630$ N

(B) $R_A = 160$ N; $R_B = 840$ N

(C) $R_A = 630$ N; $R_B = 120$ N

(D) $R_A = 840$ N; $R_B = 160$ N

2. What is most nearly the fixed-end moment at point A when a 15 N load is applied?

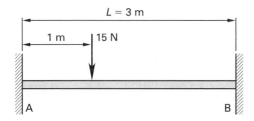

(A) 3.3 N·m

(B) 6.7 N·m

(C) 33 N·m

(D) 45 N·m

3. For the beam shown, the fixed-end moment at point A cannot exceed 2 N·m.

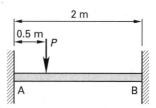

What is most nearly the maximum load that can be applied to the beam?

(A) 1.8 N

(B) 2.0 N

(C) 3.6 N

(D) 7.1 N

4. The truss shown is constructed of members having a modulus of elasticity of 200 GPa. The cross-sectional area of each member is 5 cm².

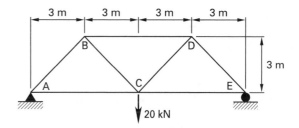

What is most nearly the horizontal displacement of the roller at location 5 due to the applied load?

(A) 0.1 cm

(B) 4.2 cm

(C) 6.0 cm

(D) 8.5 cm

SOLUTIONS

1. The total length is $L = a + b = 30 \text{ ft} + 10 \text{ ft} = 40 \text{ ft}$. Take clockwise moments to be positive.

$$\text{FEM}_{\text{AB}} = \frac{-Pab^2}{L^2} = \frac{-(1000 \text{ N})(30 \text{ m})(10 \text{ m})^2}{(40 \text{ m})^2}$$
$$= -1875 \text{ N·m}$$

$$\text{FEM}_{\text{BA}} = \frac{Pa^2b}{L^2} = \frac{(1000 \text{ N})(30 \text{ m})^2(10 \text{ m})}{(40 \text{ m})^2}$$
$$= 5625 \text{ N·m}$$

The moment equation at support A is

$$\sum M_{\text{A}} = \text{FEM}_{\text{AB}} + \text{FEM}_{\text{BA}}$$
$$+ (1000 \text{ N})(30 \text{ m}) - (40 \text{ m})R_{\text{B}} = 0$$
$$= -1875 \text{ N·m} + 5625 \text{ N·m}$$
$$+ (1000 \text{ N})(30 \text{ m}) - (40 \text{ m})R_{\text{B}} = 0$$
$$R_{\text{B}} = 843.75 \text{ N}$$

The moment equation at support 2 is

$$\sum M_{\text{B}} = \text{FEM}_{\text{AB}} + \text{FEM}_{\text{BA}} - (1000 \text{ N})(10 \text{ m})$$
$$+ (40 \text{ m})R_{\text{A}} = 0$$
$$= -1875 \text{ N·m} + 5625 \text{ N·m} - (1000 \text{ N})(10 \text{ m})$$
$$+ (40 \text{ m})R_{\text{A}} = 0$$
$$R_{\text{A}} = 156.25 \text{ N}$$

The force equilibrium requirement is

$$R_{\text{A}} + R_{\text{B}} = 1000 \text{ N}$$
$$156.25 \text{ N} + 843.75 \text{ N} = 1000 \text{ N} \quad [\text{check}]$$

$$R_{\text{A}} = 156.25 \text{ N}; \quad R_{\text{B}} = 843.75 \text{ N}$$
$$(R_{\text{A}} = 160 \text{ N}; \quad R_{\text{B}} = 840 \text{ N})$$

The answer is (B).

2.
$$a = 1 \text{ m}$$
$$b = L - a = 3 \text{ m} - 1 \text{ m} = 2 \text{ m}$$

The fixed-end moment at point A is

$$\text{FEM}_{\text{AB}} = \frac{Pab^2}{L^2} = \frac{(15 \text{ N})(1 \text{ m})(2 \text{ m})^2}{(3 \text{ m})^2}$$
$$= 6.667 \text{ N·m} \quad (6.7 \text{ N·m})$$

The answer is (B).

3. The maximum load that can be applied to the beam is

$$\text{FEM}_{\text{AB}} = \frac{Pab^2}{L^2}$$

$$P = \frac{\text{FEM}_{\text{AB}}L^2}{ab^2} = \frac{(2 \text{ N·m})(2 \text{ m})^2}{(0.5 \text{ m})(1.5 \text{ m})^2}$$
$$= 7.11 \text{ N} \quad (7.1 \text{ N})$$

The answer is (D).

4. Use the unit load method. The reaction and member forces in the truss are as shown.

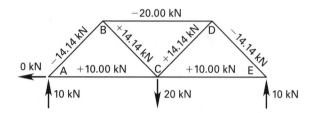

Remove the applied loads and place a unit load at member 5.

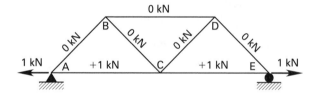

Place the results in a table.

member	L (m)	F (kN)	u (kN)	LFu (m·kN²)
AB	4.24	14.14	0	0
AC	6.00	10.00	1.00	60.00
BC	4.24	14.14	0	0
BD	6.00	20.00	0	0
CD	4.24	14.14	0	0
CE	6.00	10.00	1.00	60.00
DE	4.24	14.14	0	0
			total	120.00

The displacement at joint E is

$$f\delta = \sum \frac{SuL}{AE}$$

$$(1 \text{ kN})\delta = \frac{(120 \text{ m·kN}^2)\left(100 \frac{\text{cm}}{\text{m}}\right)^2}{(5 \text{ cm}^2)(200 \text{ GPa})\left(10^9 \frac{\text{Pa}}{\text{GPa}}\right)}$$
$$\times \left(1000 \frac{\text{N}}{\text{kN}}\right)\left(100 \frac{\text{cm}}{\text{m}}\right)$$

$$\delta = 0.12 \text{ cm} \quad (0.1 \text{ cm})$$

The answer is (A).

Statics

26 Kinematics

PRACTICE PROBLEMS

1. A particle's curvilinear motion is represented by the equation $s(t) = 20t + 4t^2 - 3t^3$. What is most nearly the particle's initial velocity?

- (A) 20 m/s
- (B) 25 m/s
- (C) 30 m/s
- (D) 32 m/s

2. A vehicle is traveling at 62 km/h when the driver sees a traffic light in an intersection 530 m ahead turn red. The light's red cycle duration is 25 s. The driver wants to enter the intersection without stopping the vehicle, just as the light turns green. If the vehicle decelerates at a constant rate of 0.35 m/s², what will be its approximate speed when the light turns green?

- (A) 31 km/h
- (B) 43 km/h
- (C) 59 km/h
- (D) 63 km/h

3. A projectile has an initial velocity of 110 m/s and a launch angle of 20° from the horizontal. The surrounding terrain is level, and air friction is to be disregarded. What is most nearly the flight time of the projectile?

- (A) 3.8 s
- (B) 7.7 s
- (C) 8.9 s
- (D) 12 s

4. A particle's position is defined by

$$\mathbf{s}(t) = 2\sin t\mathbf{i} + 4\cos t\mathbf{j} \quad [t \text{ in radians}]$$

What is most nearly the magnitude of the particle's velocity when $t = 4$ rad?

- (A) 2.6
- (B) 2.7
- (C) 3.3
- (D) 4.1

5. A roller coaster train climbs a hill with a constant gradient. During a 10 s period, the acceleration is constant at 0.4 m/s², and the average velocity of the train is 40 km/h. What is most nearly the velocity of the train after 10 s?

- (A) 9.1 m/s
- (B) 11 m/s
- (C) 13 m/s
- (D) 15 m/s

6. Choose the equation that best represents a rigid body or particle under constant acceleration.

- (A) $a = 9.81 \text{ m/s}^2 + v_0/t$
- (B) $v = a_0(t - t_0) + v_0$
- (C) $v = v_0 + \int_0^t a(t)\,dt$
- (D) $a = v_t^2/r$

7. A particle's curvilinear motion is represented by the equation $s(t) = 40t + 5t^2 - 8t^3$. What is most nearly the initial acceleration of the particle?

- (A) 2 m/s²
- (B) 3 m/s²
- (C) 8 m/s²
- (D) 10 m/s²

8. The rotor of a steam turbine is rotating at 7200 rpm when the steam supply is suddenly cut off. The rotor decelerates at a constant rate and comes to rest after 5 min. What is most nearly the angular deceleration of the rotor?

- (A) 0.40 rad/s²
- (B) 2.5 rad/s²
- (C) 5.8 rad/s²
- (D) 16 rad/s²

9. The angular position of a car traveling around a curve is described by the following function of time (in seconds).

$$\theta(t) = t^3 - 2t^2 - 4t + 10$$

What is most nearly the angular acceleration of the car at a time of 5 s?

(A) 4.0 rad/s^2

(B) 6.0 rad/s^2

(C) 26 rad/s^2

(D) 30 rad/s^2

10. A vehicle is traveling at 70 km/h when the driver sees a traffic light in the next intersection turn red. The intersection is 250 m away, and the light's red cycle duration is 15 s. What is most nearly the uniform deceleration that will put the vehicle in the intersection the moment the light turns green?

(A) 0.18 m/s^2

(B) 0.25 m/s^2

(C) 0.37 m/s^2

(D) 1.3 m/s^2

11. A projectile has an initial velocity of 85 m/s and a launch angle of 60° from the horizontal. The surrounding terrain is level, and air friction is to be disregarded. What is most nearly the horizontal distance traveled by the projectile?

(A) 80 m

(B) 400 m

(C) 640 m

(D) 1200 m

12. A particle's position is defined by

$$\mathbf{s}(t) = 15 \sin t \mathbf{i} + 8.5 \cos t \mathbf{j} \quad [t \text{ in radians}]$$

What is most nearly the magnitude of the particle's acceleration when $t = \pi$?

(A) 6.5

(B) 8.5

(C) 15

(D) 17

13. A particle's curvilinear motion is represented by the equation $s(t) = 30t - 8t^2 + 6t^3$. What is most nearly the maximum speed reached by the particle?

(A) 26 m/s

(B) 30 m/s

(C) 35 m/s

(D) 48 m/s

14. A projectile has an initial velocity of 80 m/s and a launch angle of 42° from the horizontal. The surrounding terrain is level, and air friction is to be disregarded. What is most nearly the maximum elevation achieved by the projectile?

(A) 72 m

(B) 150 m

(C) 350 m

(D) 620 m

Dynamics

SOLUTIONS

1. The initial velocity at $t = 0$ is

$$v = \frac{dr}{dt} = \frac{ds}{dt} = 20 + 8t - 9t^2$$
$$= 20 + (8)(0 \text{ s}) - (9)(0 \text{ s})^2$$
$$= 20 \text{ m/s}$$

The answer is (A).

2. The velocity after 25 s of constant deceleration is

$$v(t) = a_0(t - t_0) + v_0$$
$$= \frac{\left(-0.35 \frac{\text{m}}{\text{s}^2}\right)(25 \text{ s} - 0 \text{ s})\left(3600 \frac{\text{s}}{\text{h}}\right)}{1000 \frac{\text{m}}{\text{km}}} + 62 \frac{\text{km}}{\text{h}}$$
$$= 31 \text{ km/h}$$

The answer is (A).

3. The vertical component of velocity is zero at the apex. Calculate the time to reach the apex.

$$v_y = -gt + v_0 \sin(\theta)$$
$$0 = -\left(9.81 \frac{\text{m}}{\text{s}^2}\right)t + \left(110 \frac{\text{m}}{\text{s}}\right)\sin 20°$$
$$t = 3.84 \text{ s}$$

The projectile takes the same amount of time to return to the ground from the apex as it took to reach the apex after launch. The total flight time is

$$t_{\text{total}} = (2)(3.84 \text{ s}) = 7.67 \text{ s} \quad (7.7 \text{ s})$$

The answer is (B).

4. The velocity is

$$\mathbf{v}(t) = \frac{d\mathbf{s}(t)}{dt} = \frac{d}{dt}(2 \sin t\mathbf{i} + 4 \cos t\mathbf{j})$$
$$= 2 \cos t\mathbf{i} - 4 \sin t\mathbf{j}$$

At $t = 4$ rad,

$$\mathbf{v}(4) = 2 \cos(4 \text{ rad})\mathbf{i} - 4 \sin(4 \text{ rad})\mathbf{j}$$
$$= -1.31\mathbf{i} - (-3.03)\mathbf{j}$$
$$|\mathbf{v}(4)| = \sqrt{(-1.31\mathbf{i})^2 + (3.03\mathbf{i})^2}$$
$$= 3.3$$

The answer is (C).

5. If the train travels for 10 s at an average velocity of 40 km/h, then the distance traveled in 10 s is

$$s(t) = v_{\text{ave}}t = \frac{\left(40 \frac{\text{km}}{\text{h}}\right)\left(1000 \frac{\text{m}}{\text{km}}\right)(10 \text{ s})}{3600 \frac{\text{s}}{\text{h}}}$$
$$= 111.1 \text{ m}$$

Rearrange the equation for distance as a function of initial velocity and acceleration, and solve for the initial velocity.

$$s(t) = a_0(t - t_0)^2/2 + v_0(t - t_0) + s_0$$
$$v_0 = \frac{s(t) - s_0 - \dfrac{a_0(t - t_0)^2}{2}}{t - t_0}$$
$$= \frac{111.1 \text{ m} - 0 \text{ m} - \dfrac{\left(0.4 \frac{\text{m}}{\text{s}^2}\right)(10 \text{ s} - 0 \text{ s})^2}{2}}{10 \text{ s} - 0 \text{ s}}$$
$$= 9.11 \text{ m/s}$$

For an initial velocity of 9.11 m/s and an acceleration of 0.4 m/s² over 10 s, the final velocity after 10 s is

$$v_f = v_0 + a_0 t$$
$$= 9.11 \frac{\text{m}}{\text{s}} + \left(0.4 \frac{\text{m}}{\text{s}^2}\right)(10 \text{ s})$$
$$= 13.11 \text{ m/s} \quad (13 \text{ m/s})$$

The answer is (C).

6. Option A is an expression for acceleration that varies with time. Option C is an expression for velocity with a generalized time-varying acceleration. The expression in option D relates tangential and normal accelerations, respectively, along a curved path, to the tangential velocity. For a generalized curved path, these accelerations are not constant.

Option B is the expression for the velocity of a linear system under constant acceleration.

$$v(t) = a_0 \int dt = a_0(t - t_0) + v_0$$

The answer is (B).

7. The acceleration at $t = 0$ is

$$a = \frac{d^2r}{dt^2} = \frac{d^2s}{dt^2} = 10 - 48t$$
$$= 10 \text{ m/s}^2$$

The answer is (D).

Dynamics

8. The angular deceleration (velocity) is

$$\omega_f = \omega_0 - \alpha t$$

$$\alpha = \frac{\omega_0 - \omega_f}{t}$$

$$= \frac{\left(7200 \ \frac{\text{rev}}{\text{min}}\right)\left(2\pi \ \frac{\text{rad}}{\text{rev}}\right) - 0 \ \frac{\text{rad}}{\text{s}}}{(5 \ \text{min})\left(60 \ \frac{\text{s}}{\text{min}}\right)^2}$$

$$= 2.51 \ \text{rad/s}^2 \quad (2.5 \ \text{rad/s}^2)$$

The answer is (B).

9. The angular acceleration is

$$\alpha(t) = \frac{d^2\theta}{dt^2} = 6t - 4$$

$$\alpha(5) = (6)(5 \ \text{s}) - 4 = 26 \ \text{rad/s}^2$$

The answer is (C).

10. Rearrange the equation for the distance traveled under a constant acceleration. Let the initial distance traveled equal 0 m, and the initial time equal 0 s.

$$s(t) = a_0(t - t_0)^2/2 + v_0(t - t_0) + s_0$$

$$a_0 = \frac{(2)(-v_0(t - t_0) - s(t) - s_0)}{(t - t_0)^2}$$

$$= \frac{(2)\left(\dfrac{\left(-70 \ \frac{\text{km}}{\text{h}}\right)\left(1000 \ \frac{\text{m}}{\text{km}}\right)(15 \ \text{s} - 0 \ \text{s})}{3600 \ \frac{\text{s}}{\text{h}}} + 250 \ \text{m} - 0 \ \text{m}\right)}{(15 \ \text{s} - 0 \ \text{s})^2}$$

$$= -0.37 \ \text{m/s}^2 \quad (0.37 \ \text{m/s}^2 \ \text{deceleration})$$

The answer is (C).

11. Calculate the total flight time. The vertical component of velocity is zero at the apex.

$$v_y = -gt + v_0 \sin(\theta)$$

$$0 = -\left(9.81 \ \frac{\text{m}}{\text{s}^2}\right)t + \left(85 \ \frac{\text{m}}{\text{s}}\right)\sin 60°$$

$$t = 7.50 \ \text{s}$$

The projectile takes the same amount of time to return to the ground from the apex as it took to reach the apex after launch. The total flight time is

$$t = (2)(7.50 \ \text{s}) = 15.0 \ \text{s}$$

The horizontal distance traveled is

$$x = v_0 \cos(\theta)t + x_0$$

$$= \left(85 \ \frac{\text{m}}{\text{s}}\right)\cos 60°(15.0 \ \text{s}) + 0 \ \text{m}$$

$$= 638 \ \text{m} \quad (640 \ \text{m})$$

The answer is (C).

12. The velocity is

$$\mathbf{v}(t) = \frac{d\mathbf{s}(t)}{dt} = \frac{d}{dt}(15 \sin t\mathbf{i} + 8.5 \cos t\mathbf{j})$$

$$= 15 \cos t\mathbf{i} - 8.5 \sin t\mathbf{j}$$

$$\mathbf{a}(t) = \frac{d\mathbf{v}(t)}{dt} = -15 \sin t\mathbf{i} - 8.5 \cos t\mathbf{j}$$

$$\mathbf{a}(\pi) = -15 \sin \pi\mathbf{i} - 8.5 \cos \pi\mathbf{j}$$

$$= 0\mathbf{i} + 8.5\mathbf{j}$$

$$|\mathbf{a}(\pi)| = \sqrt{(0\mathbf{i})^2 + (8.5\mathbf{j})^2}$$

$$= 8.5$$

The answer is (B).

13. The maximum of the velocity function is found by equating the derivative of the velocity function to zero and solving for t.

$$v(t) = \frac{ds}{dt} = \frac{d}{dt}(30t - 8t^2 + 6t^3)$$

$$= 30 - 16t + 18t^2$$

$$\frac{dv}{dt} = \frac{d}{dt}(30 - 16t + 18t^2) = -16 + 36t = 0$$

$$t = 0.444 \ \text{s}$$

$$v_{\max} = 30 - 16t + 18t^2$$

$$= 30 - (16)(0.444 \ \text{s}) - (18)(0.444 \ \text{s})^2$$

$$= 26.4 \ \text{m/s} \quad (26 \ \text{m/s})$$

The answer is (A).

14. The maximum elevation is achieved when the projectile is at the apex. The vertical component of velocity is zero at the apex. Calculate the time to reach the apex.

$$v_y = -gt + v_0 \sin(\theta)$$

$$0 = -\left(9.81 \ \frac{\text{m}}{\text{s}^2}\right)t + \left(80 \ \frac{\text{m}}{\text{s}}\right)\sin 42°$$

$$t = 5.46 \ \text{s}$$

The elevation at time t is

$$y = -gt^2/2 + v_0 \sin(\theta)t + y_0$$

$$= \frac{-\left(9.81 \ \frac{\text{m}}{\text{s}^2}\right)(5.46 \ \text{s})^2}{2}$$

$$+ \left(80 \ \frac{\text{m}}{\text{s}}\right)\sin 42°(5.46 \ \text{s}) + 0 \ \text{m}$$

$$= 146 \ \text{m} \quad (150 \ \text{m})$$

The answer is (B).

Dynamics

27 Kinetics

PRACTICE PROBLEMS

1. The 52 kg block shown starts from rest at position A and slides down the inclined plane to position B. The coefficient of friction between the block and the plane is $\mu = 0.15$.

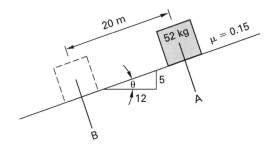

What is most nearly the velocity of the block at position B?

(A) 2.4 m/s

(B) 4.1 m/s

(C) 7.0 m/s

(D) 9.8 m/s

2. A 5 kg block begins from rest and slides down an inclined plane. After 4 s, the block has a velocity of 6 m/s. If the angle of inclination of the plane is 45°, approximately how far has the block traveled after 4 s?

(A) 1.5 m

(B) 3.0 m

(C) 6.0 m

(D) 12 m

3. The elevator in a 20-story apartment building has a mass of 1800 kg. Its maximum velocity and maximum acceleration are 2.5 m/s and 1.4 m/s², respectively. A passenger weighing 67 kg stands on a scale in the elevator as the elevator ascends at its maximum acceleration. When the elevator reaches its maximum acceleration, the scale most nearly reads

(A) 67 N

(B) 560 N

(C) 660 N

(D) 750 N

4. A rope is used to tow an 800 kg car with free-rolling wheels over a smooth, level road. The rope will break if the tension exceeds 2000 N. What is most nearly the greatest acceleration that the car can reach without breaking the rope?

(A) 1.2 m/s²

(B) 2.5 m/s²

(C) 3.8 m/s²

(D) 4.5 m/s²

5. An 8 kg block begins from rest and slides down an inclined plane. After 10 s, the block has a velocity of 15 m/s. The plane's angle of inclination is 30°. What is most nearly the coefficient of friction between the plane and the block?

(A) 0.15

(B) 0.22

(C) 0.40

(D) 0.85

6. If the sum of the forces on a particle is not equal to zero, the particle is

(A) moving with constant velocity in the direction of the resultant force

(B) accelerating in a direction opposite to the resultant force

(C) accelerating in the same direction as the resultant force

(D) moving with a constant velocity opposite to the direction of the resultant force

7. A 383 N horizontal force is applied to the 65 kg block shown. Beginning at position A, the block moves down the slope at a velocity of 12.5 m/s and comes to a complete stop at position B. The coefficient of friction between the block and the plane is $\mu = 0.22$.

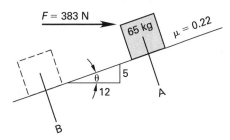

What is most nearly the distance between positions A and B?

(A) 6.1 m

(B) 9.1 m

(C) 15 m

(D) 19 m

SOLUTIONS

1. Choose a coordinate system parallel and perpendicular to the plane, as shown. Let the x-axis be positive in the direction of motion (to the left). Recognize that this is a 5-12-13 triangle. Alternatively, calculate $\sqrt{(5)^2 + (12)^2} = 13$.

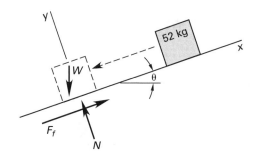

From the equations for friction and the equation for the radial component of force, the acceleration is

$$\sum F_x = ma_x$$
$$W_x - \mu N = ma_x$$
$$mg\sin\theta - \mu mg\cos\theta = ma_x$$

$$a_x = g\sin\theta - \mu g\cos\theta = g(\sin\theta - \mu\cos\theta)$$
$$= \left(9.81 \ \frac{\text{m}}{\text{s}^2}\right)\left(\frac{5}{13} - (0.15)\left(\frac{12}{13}\right)\right)$$
$$= 2.415 \ \text{m/s}^2$$

The velocity at position B is

$$v^2 = v_0^2 + 2a_x(x - x_0)$$
$$v_0 = x_0 = 0$$
$$v^2 = 2a_x x$$
$$= (2)\left(2.415 \ \frac{\text{m}}{\text{s}^2}\right)(20 \ \text{m})$$
$$= 96.6 \ \text{m}^2/\text{s}^2$$
$$v = \sqrt{96.6 \ \frac{\text{m}^2}{\text{s}^2}} = 9.83 \ \text{m/s} \quad (9.8 \ \text{m/s})$$

The answer is (D).

2. Calculate the initial acceleration.

$$v(t) = a_x t + v_0$$
$$a_x = \frac{v(t) - v_0}{t} = \frac{6 \ \frac{\text{m}}{\text{s}} - 0 \ \frac{\text{m}}{\text{s}}}{4 \ \text{s}}$$
$$= 1.5 \ \text{m/s}^2$$

Dynamics

After 4 s the block will have moved

$$x = \frac{a_x t^2}{2} + v_0 t + x_0$$

$$= \frac{\left(1.5 \ \frac{m}{s^2}\right)(4 \ s)^2}{2} + \left(0 \ \frac{m}{s}\right)(4 \ s) + 0 \ m$$

$$= 12 \ m$$

The answer is (D).

3. Use Newton's second law. The acceleration of the elevator adds to the gravitational acceleration.

$$F = ma = m(a_1 + a_2)$$

$$= (67 \ kg)\left(9.81 \ \frac{m}{s^2} + 1.4 \ \frac{m}{s^2}\right)$$

$$= 751 \ N \quad (750 \ N)$$

The answer is (D).

4. From Newton's second law, the maximum acceleration is

$$F = ma$$

$$a = \frac{F}{m} = \frac{2000 \ N}{800 \ kg}$$

$$= 2.5 \ m/s^2$$

The answer is (B).

5. Calculate the initial acceleration.

$$v_x = a_x t + v_0$$

$$a_x = \frac{v_x - v_0}{t} = \frac{15 \ \frac{m}{s} - 0 \ \frac{m}{s}}{10 \ s}$$

$$= 1.5 \ m/s^2$$

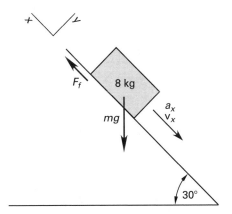

Choose a coordinate system so that the x-direction is parallel to the inclined plane. From the equations for friction, normal force, and the parallel component of force, the coefficient of friction is

$$\sum F_x = ma_x = mg_x - F_f$$

$$ma_x = mg\sin\phi - \mu mg\cos\phi$$

$$\mu = \frac{mg\sin\phi - ma_x}{mg\cos\phi}$$

$$= \frac{g\sin\phi - a_x}{g\cos\phi}$$

$$= \frac{\left(9.81 \ \frac{m}{s^2}\right)\sin 30° - 1.5 \ \frac{m}{s^2}}{\left(9.81 \ \frac{m}{s^2}\right)\cos 30°}$$

$$= 0.40$$

The answer is (C).

6. Newton's second law can be applied separately to any direction in which forces are resolved into components, including the resultant direction.

$$F_R = ma_R$$

Since force and acceleration are both vectors, and mass is a scalar, the direction of acceleration is the same as the resultant force.

$$\mathbf{a}_R = \frac{\mathbf{F}_R}{m}$$

The answer is (C).

7. Choose a coordinate system parallel and perpendicular to the plane, as shown. Let the x-axis be positive in the direction of motion (to the left). Recognize that this is a 5-12-13 triangle. Alternatively, calculate $\sqrt{(5)^2 + (12)^2} = 13$.

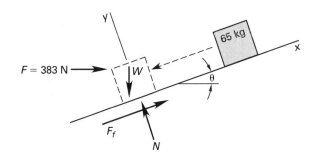

Using the sum of forces and the equations for friction and normal force, solve for the acceleration.

$$\sum F_t = ma_t$$

$$W_x - F_x - \mu N = ma_t$$

$$mg\sin\theta - F\cos\theta - \mu(mg\cos\theta + F\sin\theta) = ma_t$$

$$
\begin{aligned}
a_t &= \left(\frac{1}{m}\right)\left(mg\sin\theta - F\cos\theta - \mu(mg\cos\theta + F\sin\theta)\right) \\
&= g\left(\sin\theta - \mu g\cos\theta\right) - \frac{F}{m}\left(\cos\theta + \mu\sin\theta\right) \\
&= \left(9.81\ \frac{\text{m}}{\text{s}^2}\right)\left(\frac{5}{13} - (0.22)\left(\frac{12}{13}\right)\right) \\
&\quad - \left(\frac{383\ \text{N}}{65\ \text{kg}}\right)\left(\frac{12}{13} + (0.22)\left(\frac{5}{13}\right)\right) \\
&= -4.157\ \text{m/s}^2
\end{aligned}
$$

The velocity is

$$v^2 = v_0^2 + 2a_t(x - x_0)$$

$$v_0 = 12.5\ \text{m/s}$$

$$v = x_0 = 0$$

The distance between positions A and B is

$$
x = \frac{-v_0^2}{2a_t} = \frac{-\left(12.5\ \frac{\text{m}}{\text{s}}\right)^2}{(2)\left(-4.157\ \frac{\text{m}}{\text{s}^2}\right)}
$$

$$= 18.79\ \text{m} \quad (19\ \text{m})$$

The answer is (D).

Dynamics

28 Kinetics of Rotational Motion

PRACTICE PROBLEMS

1. A 1530 kg car is towing a 300 kg trailer. The coefficient of friction between all tires and the road is 0.80. The car and trailer are traveling at 100 km/h around a banked curve of radius 200 m. What is most nearly the necessary banking angle such that tire friction will NOT be necessary to prevent skidding?

(A) 8.0°

(B) 21°

(C) 36°

(D) 78°

2. Why does a spinning ice skater's angular velocity increase as she brings her arms in toward her body?

(A) Her mass moment of inertia is reduced.

(B) Her angular momentum is constant.

(C) Her radius of gyration is reduced.

(D) all of the above

3. A 1 m long uniform rod has a mass of 10 kg. It is pinned at one end to a frictionless pivot. What is most nearly the mass moment of inertia of the rod taken about the pivot point?

(A) 0.83 kg·m^2

(B) 2.5 kg·m^2

(C) 3.3 kg·m^2

(D) 10 kg·m^2

4. In the linkage mechanism shown, link AB rotates with an instantaneous counterclockwise angular velocity of 10 rad/s.

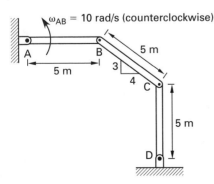

What is most nearly the instantaneous angular velocity of link BC when link AB is horizontal and link CD is vertical?

(A) 2.3 rad/s (clockwise)

(B) 3.3 rad/s (counterclockwise)

(C) 5.5 rad/s (clockwise)

(D) 13 rad/s (clockwise)

5. Two 2 kg blocks are linked as shown.

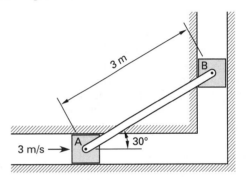

Assuming that the surfaces are frictionless, what is most nearly the velocity of block B if block A is moving at a speed of 3 m/s?

(A) 0 m/s

(B) 1.3 m/s

(C) 1.7 m/s

(D) 5.2 m/s

6. A car travels on a perfectly horizontal, unbanked circular track of radius r. The coefficient of friction between the tires and the track is 0.3. If the car's velocity is 10 m/s, what is most nearly the smallest radius the car can travel without skidding?

(A) 10 m

(B) 34 m

(C) 50 m

(D) 68 m

7. A uniform rod (AB) of length L and weight W is pinned at point C. The rod starts from rest and accelerates with an angular acceleration of $12g/7L$.

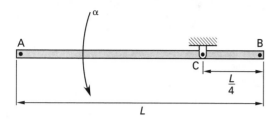

What is the instantaneous reaction at point C at the moment rotation begins?

(A) $\dfrac{W}{4}$

(B) $\dfrac{W}{3}$

(C) $\dfrac{4W}{7}$

(D) $\dfrac{7W}{12}$

8. A wheel with a 0.75 m radius has a mass of 200 kg. The wheel is pinned at its center and has a radius of gyration of 0.25 m. A rope is wrapped around the wheel and supports a hanging 100 kg block. When the wheel is released, the rope begins to unwind. What is most nearly the angular acceleration of the wheel as the block descends?

(A) 5.9 rad/s^2

(B) 6.5 rad/s^2

(C) 11 rad/s^2

(D) 14 rad/s^2

9. A car travels around an unbanked 50 m radius curve without skidding. The coefficient of friction between the tires and road is 0.3. What is most nearly the car's maximum velocity?

(A) 14 km/h

(B) 25 km/h

(C) 44 km/h

(D) 54 km/h

10. A uniform rod (AB) of length L and weight W is pinned at point C and restrained by cable OA. The cable is suddenly cut. The rod starts to rotate about point C, with point A moving down and point B moving up.

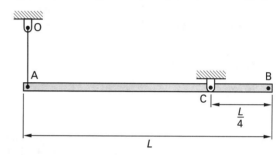

The instantaneous linear acceleration of point B is

(A) $\dfrac{3g}{16}$

(B) $\dfrac{g}{4}$

(C) $\dfrac{3g}{7}$

(D) $\dfrac{3g}{4}$

SOLUTIONS

1. The necessary superelevation angle without relying on friction is

$$\theta = \arctan \frac{v^2}{gr}$$

$$= \arctan \frac{\left(\left(100 \ \frac{km}{h}\right)\left(1000 \ \frac{m}{km}\right)\right)^2}{\left(9.81 \ \frac{m}{s^2}\right)(200 \ m)\left(3600 \ \frac{s}{h}\right)^2}$$

$$= 21.47° \quad (21°)$$

The answer is (B).

2. As the skater brings her arms in, her radius of gyration and mass moment of inertia decrease. However, in the absence of friction, her angular momentum, H, is constant.

$$\omega = \frac{H}{I}$$

Since angular velocity, ω, is inversely proportional to the mass moment of inertia, the angular velocity increases when the mass moment of inertia decreases.

The answer is (D).

3. The mass moment of inertia of the rod taken about one end is

$$I_{rod} = \frac{ML^2}{3} = \frac{(10 \ kg)(1 \ m)^2}{3}$$

$$= 3.33 \ kg{\cdot}m^2 \quad (3.3 \ kg{\cdot}m^2)$$

The answer is (C).

4. Find the instantaneous center of rotation. The absolute velocity directions at points B and C are known. The instantaneous center is located by drawing perpendiculars to these velocities, as shown. The angular velocity of any point on rigid body link BC is the same at this instant.

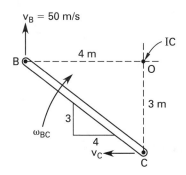

The velocity of point B is

$$v_B = AB\omega_{AB} = (5 \ m)\left(10 \ \frac{rad}{s}\right) = 50 \ m/s$$

The angular velocity of link BC is

$$\omega_{BC} = \frac{v_B}{OB} = \frac{50 \ \frac{m}{s}}{4 \ m}$$

$$= 12.5 \ rad/s \quad (13 \ rad/s) \quad [clockwise]$$

The answer is (D).

5. The instantaneous center of rotation for the slider rod assembly can be found by extending perpendiculars from the velocity vectors, as shown. Both blocks can be assumed to rotate about point C with angular velocity ω.

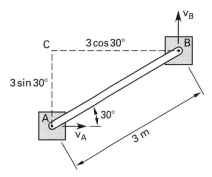

The velocity of block B is

$$\omega = \frac{v_A}{CA} = \frac{v_B}{BC}$$

$$v_B = \frac{v_A BC}{CA} = \frac{\left(3 \ \frac{m}{s}\right)(3 \ m)\cos 30°}{(3 \ m)\sin 30°}$$

$$= 5.2 \ m/s$$

The answer is (D).

6. If there is no skidding, the frictional force, F_f, will equal the centrifugal force, F_c. From the equations for centrifugal force and frictional force, the smallest possible radius is

$$F_c = \frac{mv^2}{r}$$

$$F_f = \mu N = \mu mg$$

$$\frac{mv^2}{r} = \mu mg$$

$$r = \frac{v^2}{\mu g} = \frac{\left(10 \ \frac{m}{s}\right)^2}{(0.3)\left(9.81 \ \frac{m}{s^2}\right)} = 34 \ m$$

The answer is (B).

Dynamics

7. The mass moment of inertia of the rod about its center of gravity is

$$I_{CG} = \frac{ML^2}{12} = \left(\frac{W}{g}\right)\left(\frac{L^2}{12}\right)$$

Take moments about the center of gravity of the rod. All moments due to gravitational forces will cancel. The only unbalanced force acting on the rod will be the vertical reaction force, R_C, at point C.

$$\sum M_{CG} = R_C\left(\frac{L}{4}\right) = I_{CG}\alpha_{CG}$$

$$R_C\left(\frac{L}{4}\right) = \left(\left(\frac{W}{g}\right)\left(\frac{L^2}{12}\right)\right)\left(\frac{12g}{7L}\right)$$

$$R_C = \frac{4W}{7}$$

The angular velocity is zero, so the center of the mass does not have a component of acceleration in the horizontal direction. There is no horizontal force component at point C.

The answer is (C).

8. From the equation for the radius of gyration, the mass moment of inertia of the wheel is

$$r = \sqrt{\frac{I}{m_{wheel}}}$$

$$I = m_{wheel}r^2$$

The unbalanced moment on the wheel is

$$M = FR = (mg - ma)R = mR(g - a)$$
$$= m_{block}R(g - R\alpha)$$

The acceleration is given by

$$M = I\alpha$$

$$m_{block}R(g - R\alpha) = m_{wheel}r^2\alpha$$

Combine the equations and solve.

$$\alpha = \frac{m_{block}Rg}{m_{wheel}r^2 + m_{block}R^2}$$

$$= \frac{(100 \text{ kg})(0.75 \text{ m})\left(9.81 \frac{m}{s^2}\right)}{(200 \text{ kg})(0.25 \text{ m})^2 + (100 \text{ kg})(0.75 \text{ m})^2}$$

$$= 10.7 \text{ rad/s}^2 \quad (11 \text{ rad/s}^2)$$

The answer is (C).

9. If the car does not skid, the frictional force and the centrifugal force must be equal. From the equations for centrifugal force and frictional force, the car's maximum velocity is

$$F_c = \frac{mv^2}{r}$$

$$F_f = \mu N = \mu mg$$

$$\frac{mv^2}{r} = \mu mg$$

$$v = \sqrt{\mu gr}$$

$$= \sqrt{\begin{array}{c}(0.3)\left(\dfrac{\left(9.81 \frac{m}{s^2}\right)\left(60 \frac{s}{min}\right)^2\left(60 \frac{min}{h}\right)^2}{1000 \frac{m}{km}}\right) \\ \times \left(\dfrac{50 \text{ m}}{1000 \frac{m}{km}}\right)\end{array}}$$

$$= 43.67 \text{ km/h} \quad (44 \text{ km/h})$$

The answer is (C).

10. Point C is $L/4$ from the center of gravity of the rod. The mass moment of inertia about point C is

$$I_C = I_{CG} + Md^2 = \frac{ML^2}{12} + M\left(\frac{L}{4}\right)^2 = \left(\frac{7}{48}\right)ML^2$$

The sum of moments on the rod is

$$\sum M_C = \sum Fr = \left(\frac{3W}{4}\right)\left(\frac{3L}{4}\right) - \left(\frac{W}{4}\right)\left(\frac{L}{4}\right)$$

$$= \frac{WL}{4}$$

$$= \frac{MgL}{4}$$

The angular acceleration is

$$\alpha = \frac{\sum M_C}{I_C} = \frac{\frac{MgL}{4}}{\left(\frac{7}{48}\right)ML^2} = \frac{12g}{7L}$$

The tangential acceleration of point B is

$$a_{t,B} = r\alpha = \left(\frac{L}{4}\right)\left(\frac{12g}{7L}\right) = \frac{3g}{7}$$

The answer is (C).

Dynamics

29 Energy and Work

PRACTICE PROBLEMS

1. The 40 kg mass, m, in the illustration shown is guided by a frictionless rail. The spring constant, k, is 3000 N/m. The spring is compressed sufficiently and released, such that the mass barely reaches point B.

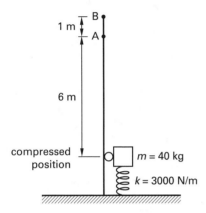

What is most nearly the initial spring compression?

(A) 0.96 m

(B) 1.3 m

(C) 1.4 m

(D) 1.8 m

2. Two balls, both of mass 2 kg, collide head on. The velocity of each ball at the time of the collision is 2 m/s. The coefficient of restitution is 0.5. Most nearly, what are the final velocities of the balls?

(A) 1 m/s and −1 m/s

(B) 2 m/s and −2 m/s

(C) 3 m/s and −3 m/s

(D) 4 m/s and −4 m/s

3. A 1500 kg car traveling at 100 km/h is towing a 250 kg trailer. The coefficient of friction between the tires and the road is 0.8 for both the car and trailer. Approximately what energy is dissipated by the brakes if the car and trailer are braked to a complete stop?

(A) 96 kJ

(B) 390 kJ

(C) 580 kJ

(D) 680 kJ

4. A 3500 kg car traveling at 65 km/h skids and hits a wall 3 s later. The coefficient of friction between the tires and the road is 0.60. What is most nearly the speed of the car when it hits the wall?

(A) 0.14 m/s

(B) 0.40 m/s

(C) 5.1 m/s

(D) 6.2 m/s

5. In the illustration shown, the 170 kg mass, m, is guided by a frictionless rail. The spring is compressed sufficiently and released, such that the mass barely reaches point B.

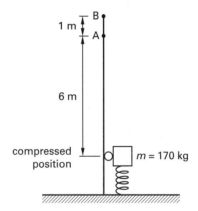

What is most nearly the kinetic energy of the mass at point A?

(A) 20 J

(B) 220 J

(C) 390 J

(D) 1700 J

6. A pickup truck is traveling forward at 25 m/s. The bed is loaded with boxes whose coefficient of friction with the bed is 0.40. What is most nearly the shortest time that the truck can be brought to a stop such that the boxes do not shift?

(A) 2.3 s

(B) 4.7 s

(C) 5.9 s

(D) 6.4 s

7. Two balls both have a mass of 8 kg and collide head on. The velocity of each ball at the time of collision is 18 m/s. The velocity of each ball decreases to 10 m/s in opposite directions after the collision. Approximately how much energy is lost in the collision?

(A) 0.57 kJ

(B) 0.91 kJ

(C) 1.8 kJ

(D) 2.3 kJ

8. The impulse-momentum principle is mostly useful for solving problems involving

(A) force, velocity, and time

(B) force, acceleration, and time

(C) velocity, acceleration, and time

(D) force, velocity, and acceleration

9. A 12 kg aluminum box is dropped from rest onto a large wooden beam. The box travels 0.2 m before contacting the beam. After impact, the box bounces 0.05 m above the beam's surface. Approximately what impulse does the beam impart on the box?

(A) 8.6 N·s

(B) 12 N·s

(C) 36 N·s

(D) 42 N·s

10. The 85 kg mass, m, shown is guided by a frictionless rail. The spring is compressed sufficiently and released, such that the mass barely reaches point B. The spring constant, k, is 1500 N/m.

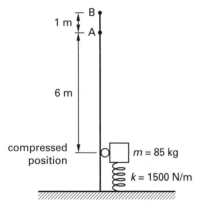

What is most nearly the velocity of the mass at point A?

(A) 3.1 m/s

(B) 4.4 m/s

(C) 9.8 m/s

(D) 20 m/s

SOLUTIONS

1. At the point just before the spring is released, all of the energy in the system is elastic potential energy; while at point B, all of the energy is potential energy due to gravity.

$$\frac{kx^2}{2} = mgh$$

$$x = \sqrt{\frac{2mgh}{k}}$$

$$= \sqrt{\frac{(2)(40 \text{ kg})\left(9.81 \frac{\text{m}}{\text{s}^2}\right)(6 \text{ m} + 1 \text{ m})}{3000 \frac{\text{N}}{\text{m}}}}$$

$$= 1.35 \text{ m} \quad (1.4 \text{ m})$$

The answer is (C).

2. Since the two velocities are in opposite directions, let the velocity of one ball, v_1, equal 2 m/s and the velocity of the other ball, v_2, equal -2 m/s.

From the definition of the coefficient of restitution,

$$e = \frac{v_2' - v_1'}{v_1 - v_2}$$

$$v_2' - v_1' = e(v_1 - v_2)$$

$$= (0.5)\left(2 \frac{\text{m}}{\text{s}} - \left(-2 \frac{\text{m}}{\text{s}}\right)\right)$$

$$= 2 \text{ m/s} \qquad \text{[Eq. I]}$$

From the conservation of momentum,

$$m_1v_1 + m_2v_2 = m_1v_1' + m_2v_2'$$

But, $m_1 = m_2$.

$$v_1 + v_2 = v_1' + v_2'$$

Since $v_1 = 2$ m/s and $v_2 = -2$ m/s,

$$v_1 + v_2 = 2 \frac{\text{m}}{\text{s}} + \left(-2 \frac{\text{m}}{\text{s}}\right) = 0$$

So,

$$v_1' + v_2' = 0 \qquad \text{[Eq. II]}$$

Solve Eq. I and Eq. II simultaneously by adding them.

$$v_1' = -1 \text{ m/s}$$

$$v_2' = 1 \text{ m/s}$$

The answer is (A).

3. The original velocity of the car and trailer is

$$v = \frac{\left(100 \ \frac{\text{km}}{\text{h}}\right)\left(1000 \ \frac{\text{m}}{\text{km}}\right)}{\left(60 \ \frac{\text{s}}{\text{min}}\right)\left(60 \ \frac{\text{min}}{\text{h}}\right)} = 27.78 \text{ m/s}$$

Since the final velocity is zero, the energy dissipated is the original kinetic energy.

$$T = \frac{mv^2}{2} = \frac{(1500 \text{ kg} + 250 \text{ kg})\left(27.78 \ \frac{\text{m}}{\text{s}}\right)^2}{2}$$
$$= 675\,154 \text{ J} \quad (680 \text{ kJ})$$

The answer is (D).

4. The frictional force (negative because it opposes motion) decelerating the car is

$$F_f = -\mu N = -\mu mg$$
$$= -(0.60)(3500 \text{ kg})\left(9.81 \ \frac{\text{m}}{\text{s}^2}\right)$$
$$= -20\,601 \text{ N}$$

Use the impulse-momentum principle.

$$F_f(t_1 - t_2) = m(v_1 - v_2)$$
$$v_2 = v_1 - \frac{F_f(t_1 - t_2)}{m}$$
$$= \frac{\left(65 \ \frac{\text{km}}{\text{h}}\right)\left(1000 \ \frac{\text{m}}{\text{km}}\right)}{\left(60 \ \frac{\text{s}}{\text{min}}\right)\left(60 \ \frac{\text{min}}{\text{h}}\right)}$$
$$- \frac{(-20\,601 \text{ N})(0 \text{ s} - 3 \text{ s})}{3500 \text{ kg}}$$
$$= 0.3976 \text{ m/s} \quad (0.40 \text{ m/s})$$

The answer is (B).

5. At point A, the energy of the mass is a combination of kinetic and gravitational potential energies. The total energy of the system is constant, and the kinetic energy at B is 0.

$$E_A = E_B$$
$$U_A + T_A = U_B$$
$$mgh + \frac{mv^2}{2} = mg(h + 1 \text{ m})$$
$$T_A = mg(h + 1 \text{ m}) - mgh$$
$$= mg(1 \text{ m})$$
$$= (170 \text{ kg})\left(9.81 \ \frac{\text{m}}{\text{s}^2}\right)(1 \text{ m})$$
$$= 1670 \text{ J} \quad (1700 \text{ J})$$

The answer is (D).

6. The frictional force is the only force preventing the boxes from shifting. The forces on each box are its weight, the normal force, and the frictional force. The normal force on each box is equal to the box weight.

$$N = W = mg$$

The frictional force is

$$F_f = \mu N = \mu mg$$

Use the impulse-momentum principle. $v_2 = 0$. The frictional force is opposite of the direction of motion, so it is negative.

$$\text{Imp} = \Delta p$$
$$F_f \Delta t = m \Delta v$$
$$\Delta t = \frac{m(v_2 - v_1)}{F_f} = \frac{-mv_1}{-\mu mg} = \frac{v_1}{\mu g}$$
$$= \frac{25 \ \frac{\text{m}}{\text{s}}}{(0.40)\left(9.81 \ \frac{\text{m}}{\text{s}^2}\right)}$$
$$= 6.37 \text{ s} \quad (6.4 \text{ s})$$

The answer is (D).

7. Each ball possesses kinetic energy before and after the collision. The velocity of each ball is reduced from $|18 \text{ m/s}|$ to $|10 \text{ m/s}|$.

$$\Delta T = T_2 - T_1 = (2)\left(\frac{m(v_2^2 - v_1^2)}{2}\right)$$
$$= (2)\left(\frac{(8 \text{ kg})\left(\left(18 \ \frac{\text{m}}{\text{s}}\right)^2 - \left(10 \ \frac{\text{m}}{\text{s}}\right)^2\right)}{2}\right)$$
$$= 1792 \text{ J} \quad (1.8 \text{ kJ})$$

The answer is (C).

8. Impulse is calculated from force and time. Momentum is calculated from mass and velocity. The impulse-momentum principle is useful in solving problems involving force, time, velocity, and mass.

The answer is (A).

Dynamics

9. Initially, the box has potential energy only. (This takes the beam's upper surface as the reference plane.) When the box reaches the beam, all of the potential energy will have been converted to kinetic energy.

$$mgh_1 = \frac{mv_1^2}{2}$$

$$v_1 = \sqrt{2gh_1}$$

$$= \sqrt{(2)\left(9.81 \ \frac{m}{s^2}\right)(0.2 \ m)}$$

$$= 1.98 \ m/s \quad [\text{downward}]$$

When the box rebounds to its highest point, all of its remaining energy will be potential energy once again.

$$mgh_2 = \frac{mv_2^2}{2}$$

$$v_2 = \sqrt{2gh_2}$$

$$= \sqrt{(2)\left(9.81 \ \frac{m}{s^2}\right)(0.05 \ m)}$$

$$= 0.99 \ m/s \quad [\text{upward}]$$

Use the impulse-momentum principle. (Downward is taken as the positive velocity direction.)

$$\text{Imp} = \Delta p = m(v_1 - v_2)$$

$$= (12 \ kg)\left(1.98 \ \frac{m}{s} - \left(-0.99 \ \frac{m}{s}\right)\right)$$

$$= 35.66 \ N \cdot s \quad (36 \ N \cdot s)$$

The answer is (C).

10. At point A, the energy of the mass is a combination of kinetic and gravitational potential energies. The total energy of the system is constant, and the kinetic energy at B is 0.

$$E_A = E_B$$

$$U_A + T_A = U_B$$

$$mgh + \frac{mv^2}{2} = mg(h + 1 \ m)$$

$$T_A = mg(h + 1 \ m) - mgh$$

$$= mg(1 \ m)$$

$$= (85 \ kg)\left(9.81 \ \frac{m}{s^2}\right)(1 \ m)$$

$$= 833.9 \ J$$

Therefore, the velocity of the mass at point A is

$$T_A = \frac{mv^2}{2} = 833.9 \ J$$

$$v = \sqrt{\frac{2T_A}{m}}$$

$$= \sqrt{\frac{(2)(833.9 \ J)}{85 \ kg}}$$

$$= 4.43 \ m/s \quad (4.4 \ m/s)$$

The answer is (B).

Dynamics

PRACTICE PROBLEMS

1. The element is subjected to the plane stress condition shown.

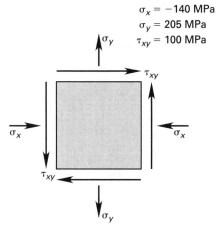

$$\sigma_x = -140 \text{ MPa}$$
$$\sigma_y = 205 \text{ MPa}$$
$$\tau_{xy} = 100 \text{ MPa}$$

What is the maximum shear stress?

(A) 100 MPa

(B) 160 MPa

(C) 200 MPa

(D) 210 MPa

2. A plane element in a body is subjected to a normal tensile stress in the x-direction of 84 MPa, as well as shear stresses of 28 MPa, as shown.

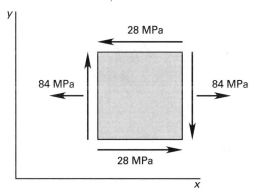

Most nearly, what are the principal stresses?

(A) 70 MPa; 14 MPa

(B) 84 MPa; 28 MPa

(C) 92 MPa; −8.5 MPa

(D) 112 MPa; −28 MPa

3. What is most nearly the lateral strain, ε_y, of the steel specimen shown if $F_x = 3000$ kN, $E = 193$ GPa, and $\nu = 0.29$?

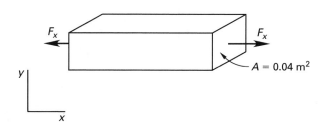

(A) -4.0×10^{-4}

(B) -1.1×10^{-4}

(C) 1.0×10^{-4}

(D) 4.0×10^{-4}

4. The elements are subjected to the plane stress condition shown. The maximum shear stress is 109.2 MPa.

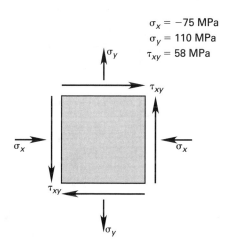

$$\sigma_x = -75 \text{ MPa}$$
$$\sigma_y = 110 \text{ MPa}$$
$$\tau_{xy} = 58 \text{ MPa}$$

What are the orientations of the stress planes (relative to the x-axis)?

(A) −74°; 15°

(B) −58°; 32°

(C) −32°; 58°

(D) −16°; 74°

Mechanics of Materials

5. What is most nearly the elongation of the aluminum bar (cross section of 3 cm × 3 cm) shown when loaded to its yield point? The modulus of elasticity is 69 GPa, and the yield strength in tension is 255 MPa. Neglect the weight of the bar.

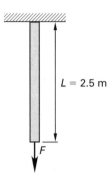

$L = 2.5$ m

F

(A) 3.3 mm

(B) 9.3 mm

(C) 12 mm

(D) 15 mm

6. The column shown has a cross-sectional area of 13 m^2.

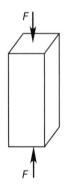

F

F

What is the approximate maximum load if the compressive stress cannot exceed 9.6 kPa?

(A) 120 kN

(B) 122 kN

(C) 125 kN

(D) 130 kN

7. The element is subjected to the plane stress condition shown. The maximum shear stress is 300 MPa.

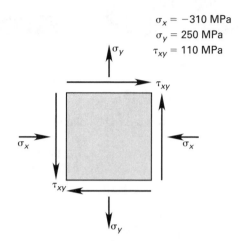

$\sigma_x = -310$ MPa
$\sigma_y = 250$ MPa
$\tau_{xy} = 110$ MPa

The principal stresses are most nearly

(A) 250 MPa; −310 MPa

(B) 270 MPa; −330 MPa

(C) 330 MPa; −270 MPa

(D) 310 MPa; −250 MPa

8. Given a shear stress of $\tau_{xy} = 35$ MPa and a shear modulus of $G = 75$ GPa, the shear strain is most nearly

(A) 2.5×10^{-5} rad

(B) 4.7×10^{-4} rad

(C) 5.5×10^{-4} rad

(D) 8.3×10^{-4} rad

9. Which of the following could be the Poisson ratio of a material?

(A) 0.35

(B) 0.52

(C) 0.55

(D) 0.60

10. A plane element in a body is subjected to the stresses shown.

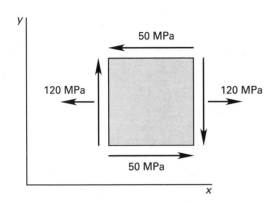

What is most nearly the maximum shear stress?

(A) 50 MPa

(B) 64 MPa

(C) 72 MPa

(D) 78 MPa

SOLUTIONS

1. There are two methods for solving the problem. The first method is to use the equation for $\tau_{\max}$; the second method is to draw Mohr's circle.

Solving by the equation for $\tau_{\max}$,

$$\tau_{\max} = \pm \sqrt{\left(\frac{\sigma_x - \sigma_y}{2}\right)^2 + \tau_{xy}^2}$$

$$= \sqrt{\left(\frac{-140 \text{ MPa} - 205 \text{ MPa}}{2}\right)^2 + (100 \text{ MPa})^2}$$

$$= 199.4 \text{ MPa} \quad (200 \text{ MPa})$$

Solving by Mohr's circle,

step 1:
$$\sigma_x = -140 \text{ MPa}$$
$$\sigma_y = 205 \text{ MPa}$$
$$\tau_{xy} = 100 \text{ MPa}$$

step 2: Draw σ-τ axes.

step 3: The circle center is

$$C = \tfrac{1}{2}(\sigma_x + \sigma_y)$$
$$= \left(\tfrac{1}{2}\right)(-140 \text{ MPa} + 205 \text{ MPa})$$
$$= 32.5 \text{ MPa}$$

step 4: Plot the points $(-140 \text{ MPa}, -100 \text{ MPa})$ and $(205 \text{ MPa}, 100 \text{ MPa})$.

step 5: Draw the diameter of the circle.

step 6: Draw the circle.

step 7: Find the radius of the circle.

step 8: Maximum shear stress is at the top of the circle, $\tau_{\max} = 199.4 \text{ MPa}$ (200 MPa).

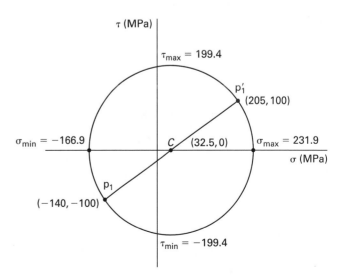

The answer is (C).

2. τ_{xy} is negative according to the standard sign convention.

$$\sigma_{\text{max,min}} = \tfrac{1}{2}(\sigma_x + \sigma_y) \pm \sqrt{\left(\frac{\sigma_x - \sigma_y}{2}\right)^2 + \tau_{xy}^2}$$

$$= \left(\tfrac{1}{2}\right)(84\text{ MPa} + 0\text{ MPa})$$

$$\pm \sqrt{\begin{array}{l}\left(\dfrac{84\text{ MPa} - 0\text{ MPa}}{2}\right)^2 \\ + (-28\text{ MPa})^2\end{array}}$$

$$= 42\text{ MPa} \pm 50.478\text{ MPa}$$

$$= 92.478\text{ MPa};\ -8.478\text{ MPa}$$

$$(92\text{ MPa};\ -8.5\text{ MPa})$$

The answer is (C).

3. From Hooke's law and the equation for axial stress,

$$\varepsilon_x = \frac{\sigma_x}{E} = \frac{F_x}{EA} = \frac{(3000\text{ kN})\left(1000\ \dfrac{\text{N}}{\text{kN}}\right)}{(193\text{ GPa})\left(10^9\ \dfrac{\text{Pa}}{\text{GPa}}\right)(0.04\text{ m}^2)}$$

$$= 3.89 \times 10^{-4}$$

Use Poisson's ratio.

$$\varepsilon_y = -\nu\varepsilon_x = (-0.29)(3.89 \times 10^{-4})$$

$$= -1.13 \times 10^{-4}\quad (-1.1 \times 10^{-4})$$

The answer is (B).

4. Calculate the angles, θ, of the stress planes.

$$\theta = \tfrac{1}{2}\arctan\frac{2\tau_{xy}}{\sigma_x - \sigma_y}$$

$$= \tfrac{1}{2}\arctan\frac{(2)(58\text{ MPa})}{-75\text{ MPa} - 110\text{ MPa}}$$

$$= -16.04°;\ 73.96°\quad (-16°;\ 74°)$$

Alternatively, the orientations can be found graphically from Mohr's circle.

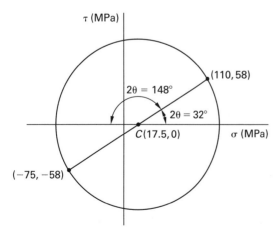

The answer is (D).

5. From Hooke's law, the axial strain is

$$\varepsilon = \frac{\sigma}{E} = \frac{(255\text{ MPa})\left(10^6\ \dfrac{\text{Pa}}{\text{MPa}}\right)}{(69\text{ GPa})\left(10^9\ \dfrac{\text{Pa}}{\text{GPa}}\right)} = 0.0037$$

The elongation is

$$\delta = \varepsilon L = (0.0037)(2.5\text{ m}) = 0.00925\text{ m}\quad (9.3\text{ mm})$$

The answer is (B).

6. The maximum force is

$$F_{\text{max}} = S_a A = (9.6\text{ kPa})(13\text{ m}^2)$$

$$= 124.8\text{ kN}\quad (125\text{ kN})$$

The answer is (C).

7. The principal stresses are

$$\sigma_{\text{max}}, \sigma_{\text{min}} = \tfrac{1}{2}(\sigma_x + \sigma_y) \pm \tau_{\text{max}}$$

$$= \left(\tfrac{1}{2}\right)(-310\text{ MPa} + 250\text{ MPa}) \pm 300\text{ MPa}$$

$$= -30\text{ MPa} \pm 300\text{ MPa}$$

$$\sigma_{\text{max}} = 270\text{ MPa}$$

$$\sigma_{\text{min}} = -330\text{ MPa}$$

The answer is (B).

8. Use Hooke's law for shear.

$$\gamma = \frac{\tau_{xy}}{G} = \frac{(35\text{ MPa})\left(10^6\ \dfrac{\text{Pa}}{\text{MPa}}\right)}{(75\text{ GPa})\left(10^9\ \dfrac{\text{Pa}}{\text{GPa}}\right)}$$

$$= 4.67 \times 10^{-4}\text{ rad}\quad (4.7 \times 10^{-4}\text{ rad})$$

The answer is (B).

9. The Poisson ratio is almost always in the range $0 < \nu < 0.5$. Option A (0.35) is the only answer that satisfies this condition.

The answer is (A).

10. The maximum shear stress is

$$\tau_{\text{max}} = \sqrt{\left(\frac{\sigma_x - \sigma_y}{2}\right)^2 + \tau_{xy}^2}$$

$$= \sqrt{\left(\frac{120\text{ MPa} - 0\text{ MPa}}{2}\right)^2 + (-50\text{ MPa})^2}$$

$$= 78.10\text{ MPa}\quad (78\text{ MPa})$$

The answer is (D).

31 Thermal, Hoop, and Torsional Stress

PRACTICE PROBLEMS

1. The maximum torque on a 0.15 m diameter solid shaft is 13 500 N·m. What is most nearly the maximum shear stress in the shaft?

(A) 20 MPa

(B) 23 MPa

(C) 28 MPa

(D) 34 MPa

2. The unrestrained glass window shown is subjected to a temperature change from 0°C to 50°C. The coefficient of thermal expansion for the glass is 8.8×10^{-6} 1/°C.

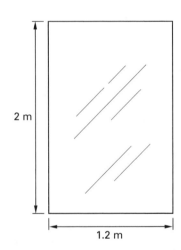

What is most nearly the change in area of the glass?

(A) 0.00040 m^2

(B) 0.0013 m^2

(C) 0.0021 m^2

(D) 0.0028 m^2

3. The cylindrical steel tank shown is 3.5 m in diameter, 5 m high, and filled with a brine solution. Brine has a density of 1198 kg/m^3. The thickness of the steel shell is 12.5 mm. Neglect the weight of the tank.

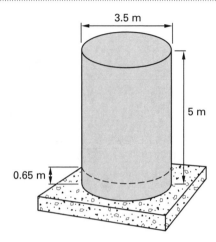

What is the approximate hoop stress in the steel 0.65 m above the rigid concrete pad?

(A) 1.2 MPa

(B) 1.4 MPa

(C) 7.2 MPa

(D) 11 MPa

4. A steel shaft is shown. The shear modulus is 80 GPa.

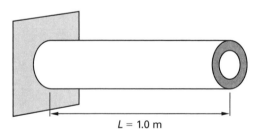

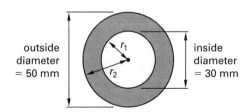

Most nearly, what torque should be applied to the end of the shaft in order to produce a twist of 1.5°?

(A) 420 N·m

(B) 560 N·m

(C) 830 N·m

(D) 1100 N·m

5. For the shaft shown, the shear stress is not to exceed 110 MPa.

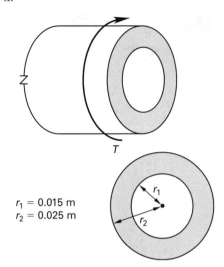

$r_1 = 0.015$ m
$r_2 = 0.025$ m

What is most nearly the largest torque that can be applied?

(A) 1700 N·m

(B) 1900 N·m

(C) 2300 N·m

(D) 3400 N·m

6. An aluminum (shear modulus $= 2.8 \times 10^{10}$ Pa) rod is 25 mm in diameter and 50 cm long. One end is rigidly fixed to a support. Most nearly, what torque must be applied at the free end to twist the rod 4.5° about its longitudinal axis?

(A) 26 N·m

(B) 84 N·m

(C) 110 N·m

(D) 170 N·m

7. A circular bar at 10°C is constrained by rigid concrete walls at both ends. The bar is 1000 mm long and has a cross-sectional area of 2600 mm^2.

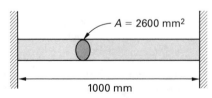

$A = 2600$ mm^2

1000 mm

E = modulus of elasticity
  = 200 GPa
α = coefficient of thermal expansion
  = 9.4×10^{-6} 1/°C

What is most nearly the axial force in the bar if the temperature is raised to 40°C?

(A) 92 kN

(B) 110 kN

(C) 130 kN

(D) 150 kN

8. A 3 m diameter bar experiences opposing torques of 280 N·m at each end.

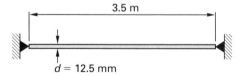

What is most nearly the maximum shear stress in the bar?

(A) 2.2 Pa

(B) 31 Pa

(C) 42 Pa

(D) 53 Pa

9. A 12.5 mm diameter steel rod is pinned between two rigid walls. The rod is initially unstressed. The rod's temperature subsequently increases 50°C. The rod is adequately stiffened and supported such that buckling does not occur. The coefficient of linear thermal expansion for steel is 11.7×10^{-6} 1/°C. The modulus of elasticity for steel is 210 GPa.

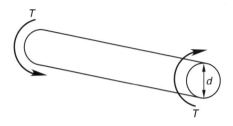

3.5 m

$d = 12.5$ mm

What is the approximate axial force in the rod?

(A) 2.8 kN

(B) 15 kN

(C) 19 kN

(D) 58 kN

10. 10 km of steel railroad track are placed when the temperature is 20°C. The linear coefficient of thermal expansion for the rails is 11×10^{-6} 1/°C. The track is free to slide forward. Most nearly, how far apart will the ends of the track be when the temperature reaches 50°C?

(A) 10.0009 km

(B) 10.0027 km

(C) 10.0033 km

(D) 10.0118 km

11. A deep-submersible diving bell has a cylindrical pressure hull with an outside diameter of 3.5 m and a wall thickness of 15 cm constructed from a ductile material. The hull is expected to experience an external pressure of 50 MPa. The hull should be designed as a

(A) thin-walled pressure vessel using the outer radius in the stress calculations

(B) thin-walled pressure vessel using the logarithmic mean area in stress calculations

(C) thin-walled pressure vessel using factors of safety of at least 4 for ductile materials and at least 8 for brittle components such as viewing ports

(D) thick-walled pressure vessel

12. A cantilever horizontal hollow tube is acted upon by a vertical force and a torque at its free end.

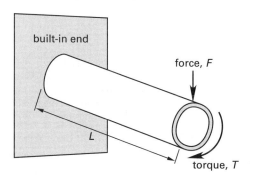

Where is the maximum stress in the cylinder?

(A) at the upper surface at midlength ($L/2$)

(B) at the lower surface at the built-in end

(C) at the upper surface at the built-in end

(D) at both the upper and lower surfaces at the built-in end

13. One end of the hollow aluminum shaft is fixed, and the other end is connected to a gear with an outside diameter of 40 cm as shown. The gear is subjected to a tangential gear force of 45 kN. The shear modulus of the aluminum is 2.8×10^{10} Pa.

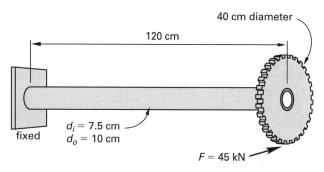

What are most nearly the maximum angle of twist and the shear stress in the shaft?

(A) 0.016 rad, 14 MPa

(B) 0.025 rad, 220 MPa

(C) 0.057 rad, 67 MPa

(D) 0.25 rad, 200 MPa

14. A compressed gas cylinder for use in a laboratory has an internal gage pressure of 8 MPa at the time of delivery. The outside diameter of the cylinder is 25 cm. If the steel has an allowable stress of 90 MPa, what is the required thickness of the wall?

(A) 0.69 cm

(B) 0.95 cm

(C) 1.1 cm

(D) 1.9 cm

SOLUTIONS

1. The polar moment of inertia is

$$J = \frac{\pi r^4}{2} = \left(\frac{\pi}{2}\right)\left(\frac{0.15 \text{ m}}{2}\right)^4$$
$$= 4.97 \times 10^{-5} \text{ m}^4$$

The shear stress is

$$\tau = \frac{Tr}{J} = \frac{(13\,500 \text{ N·m})\left(\dfrac{0.15 \text{ m}}{2}\right)}{4.97 \times 10^{-5} \text{ m}^4}$$
$$= 20.37 \times 10^6 \text{ Pa} \quad (20 \text{ MPa})$$

The answer is (A).

2. Changes in temperature affect each linear dimension.

$$\delta_{\text{width}} = \alpha L(T - T_o)$$
$$= \left(8.8 \times 10^{-6} \frac{1}{°\text{C}}\right)(1.2 \text{ m})(50°\text{C} - 0°\text{C})$$
$$= 0.000528 \text{ m}$$
$$\delta_{\text{height}} = \left(8.8 \times 10^{-6} \frac{1}{°\text{C}}\right)(2 \text{ m})(50°\text{C} - 0°\text{C})$$
$$= 0.00088 \text{ m}$$

$$A_{\text{initial}} = (2 \text{ m})(1.2 \text{ m}) = 2.4 \text{ m}^2$$
$$A_{\text{final}} = (2 \text{ m} + 0.00088 \text{ m})$$
$$\times (1.2 \text{ m} + 0.000528 \text{ m})$$
$$= 2.40211 \text{ m}^2$$
$$\Delta A = A_{\text{final}} - A_{\text{initial}}$$
$$= 2.40211 \text{ m}^2 - 2.4 \text{ m}^2$$
$$= 0.00211 \text{ m}^2 \quad (0.0021 \text{ m}^2)$$

Alternative Solution

The area coefficient of thermal expansion is, for all practical purposes, equal to 2α.

The change in area is

$$\Delta A = 2\alpha A_o \Delta T$$
$$= (2)\left(8.8 \times 10^{-6} \frac{1}{°\text{C}}\right)(2.4 \text{ m}^2)(50°\text{C} - 0°\text{C})$$
$$= 0.00211 \text{ m}^2 \quad (0.0021 \text{ m}^2)$$

The answer is (C).

3. Determine whether the tank is thin-walled or thick-walled.

$$\frac{t}{r} = \frac{12.5 \text{ mm}}{\left(\dfrac{3.5 \text{ m}}{2}\right)\left(1000 \dfrac{\text{mm}}{\text{m}}\right)} = 0.007 < 0.1$$

Use formulas for thin-walled cylindrical tanks. The pressure is

$$p = \rho g h$$
$$= \left(1198 \frac{\text{kg}}{\text{m}^3}\right)\left(9.81 \frac{\text{m}}{\text{s}^2}\right)(5 \text{ m} - 0.65 \text{ m})$$
$$= 51\,123 \text{ Pa}$$

The hoop stress is

$$\sigma_t = \frac{pd}{2t} = \frac{(51\,123 \text{ Pa})(3.5 \text{ m})}{(2)\left(\dfrac{12.5 \text{ mm}}{1000 \dfrac{\text{mm}}{\text{m}}}\right)}$$
$$= 7.157 \times 10^6 \text{ Pa} \quad (7.2 \text{ MPa})$$

The answer is (C).

4. Convert the twist angle to radians.

$$\phi = (1.5°)\left(\frac{2\pi \text{ rad}}{360°}\right) = 0.026 \text{ rad}$$

Calculate the polar moment of inertia, J.

$$r_1 = 15 \text{ mm} \quad (0.015 \text{ m})$$
$$r_2 = 25 \text{ mm} \quad (0.025 \text{ m})$$
$$J = \frac{\pi}{2}(r_o^4 - r_i^4) = \left(\frac{\pi}{2}\right)\left((0.025 \text{ m})^4 - (0.015 \text{ m})^4\right)$$
$$= 5.34 \times 10^{-7} \text{ m}^4$$

The torque is

$$T = \frac{\phi G J}{L}$$
$$= \frac{(0.026 \text{ rad})(80 \text{ GPa})\left(10^9 \dfrac{\text{Pa}}{\text{GPa}}\right)}{1.0 \text{ m}} \times (5.34 \times 10^{-7} \text{ m}^4)$$
$$= 1119 \text{ N·m} \quad (1100 \text{ N·m})$$

The answer is (D).

5. Since the shear stress is largest at the outer diameter, the maximum torque is found using this radius. For an annular region,

$$J = \frac{\pi}{2}(r_o^4 - r_i^4) = \left(\frac{\pi}{2}\right)\left((0.025 \text{ m})^4 - (0.015 \text{ m})^4\right)$$
$$= 5.34 \times 10^{-7} \text{ m}^4$$

The torque is

$$T_{\max} = \frac{\tau J}{r_2} = \frac{(110 \text{ MPa})\left(10^6 \, \frac{\text{Pa}}{\text{MPa}}\right)(5.34 \times 10^{-7} \text{ m}^4)}{0.025 \text{ m}}$$

$$= 2349.9 \text{ N·m} \quad (2300 \text{ N·m})$$

The answer is (C).

6. Convert degrees to radians.

$$\phi = (4.5°)\left(\frac{2\pi \text{ rad}}{360°}\right)$$

$$= 7.854 \times 10^{-2} \text{ rad}$$

The polar moment of inertia is

$$J = \frac{\pi r^4}{2} = \left(\frac{\pi}{2}\right)\left(\frac{25 \text{ mm}}{(2)\left(1000 \, \frac{\text{mm}}{\text{m}}\right)}\right)^4$$

$$= 3.83 \times 10^{-8} \text{ m}^4$$

Rearrange the twist angle equation to solve for torque.

$$T = \frac{\phi G J}{L}$$

$$= \frac{(7.854 \times 10^{-2} \text{ rad})(2.8 \times 10^{10} \text{ Pa})(3.83 \times 10^{-8} \text{ m}^4)}{\frac{50 \text{ cm}}{100 \, \frac{\text{cm}}{\text{m}}}}$$

$$= 168.7 \text{ N·m} \quad (170 \text{ N·m})$$

The answer is (D).

7. The elongation due to temperature change is

$$\delta = \alpha L(T_2 - T_1)$$

$$= \left(9.4 \times 10^{-6} \, \frac{1}{°\text{C}}\right)(1000 \text{ mm})(40°\text{C} - 10°\text{C})$$

$$= 0.282 \text{ mm}$$

Rearrange the elongation equation to solve for force.

$$F = \frac{\delta E A}{L}$$

$$= \frac{(0.282 \text{ mm})(200 \text{ GPa})\left(10^6 \, \frac{\text{kPa}}{\text{GPa}}\right)(2600 \text{ mm}^2)}{(1 \text{ m})\left(1000 \, \frac{\text{mm}}{\text{m}}\right)^3}$$

$$= 146.6 \text{ kN} \quad (150 \text{ kN})$$

The answer is (D).

8. Maximum shear stress occurs at the outer surface. The shear is

$$\tau = \frac{Tr}{J} = \frac{T\left(\frac{d}{2}\right)}{\frac{\pi}{32}d^4} = \frac{(280 \text{ N·m})\left(\frac{3 \text{ m}}{2}\right)}{\left(\frac{\pi}{32}\right)(3 \text{ m})^4}$$

$$= 52.8 \text{ Pa} \quad (53 \text{ Pa})$$

The answer is (D).

9. The thermal strain is

$$\varepsilon_t = \alpha \Delta T = \left(11.7 \times 10^{-6} \, \frac{1}{°\text{C}}\right)(50°\text{C})$$

$$= 0.000585 \text{ m/m}$$

The thermal stress is

$$\sigma_t = E\varepsilon_t = (210 \text{ GPa})\left(10^9 \, \frac{\text{Pa}}{\text{GPa}}\right)\left(0.000585 \, \frac{\text{m}}{\text{m}}\right)$$

$$= 1.2285 \times 10^8 \text{ Pa}$$

(This is less than the yield strength of steel.)
The compressive force in the rod is

$$F = \sigma A$$

$$= (1.2285 \times 10^8 \text{ Pa})\pi\left(\frac{12.5 \text{ mm}}{(2)\left(1000 \, \frac{\text{mm}}{\text{m}}\right)}\right)^2$$

$$= 15\,076 \text{ N} \quad (15 \text{ kN})$$

The answer is (B).

10. The total change in length is

$$\delta_t = \alpha L_{\text{initial}}(T - T_o)$$

$$= \left(11 \times 10^{-6} \, \frac{1}{°\text{C}}\right)(10 \text{ km})(50°\text{C} - 20°\text{C})$$

$$= 0.0033 \text{ km}$$

Add the change in length to the initial length.

$$L = L_{\text{initial}} + \delta_t$$

$$= 10 \text{ km} + 0.0033 \text{ km}$$

$$= 10.0033 \text{ km}$$

The answer is (C).

11. Tanks under external pressure fail by buckling (i.e., collapse), not by yielding. They should not be designed using the simplistic formulas commonly used for thin-walled tanks under internal pressure.

The answer is (D).

12. The torsional shear stress is maximum at the outer surface and is the same everywhere on the tube. The maximum moment occurs at the built-in end, tensile at the upper surface and compressive at the lower surface. The absolute value of the combined stress at the upper and lower surfaces at the built-in end will be the same.

The answer is (D).

13. Calculate the torque.

$$T = rF = \left(\frac{40 \text{ cm}}{(2)\left(100 \ \frac{\text{cm}}{\text{m}}\right)} \right) (45 \text{ kN})\left(1000 \ \frac{\text{N}}{\text{kN}}\right)$$
$$= 9000 \text{ N·m}$$

The polar moment of inertia is

$$J = \frac{\pi}{2}\left(r_o^4 - r_i^4\right)$$
$$= \left(\frac{\pi}{2}\right)\left(\left(\frac{10 \text{ cm}}{(2)\left(100 \ \frac{\text{cm}}{\text{m}}\right)} \right)^4 - \left(\frac{7.5 \text{ cm}}{(2)\left(100 \ \frac{\text{cm}}{\text{m}}\right)} \right)^4 \right)$$
$$= 6.71 \times 10^{-6} \text{ m}^4$$

Find the angle of twist.

$$\phi = \frac{TL}{GJ} = \frac{(9000 \text{ N·m})(120 \text{ cm})}{\left(2.8 \times 10^{10} \text{ Pa}\right)\left(6.71 \times 10^{-6} \text{ m}^4\right)\left(100 \ \frac{\text{cm}}{\text{m}}\right)}$$
$$= 0.057 \text{ rad}$$

Find the shear stress in the shaft.

$$\tau = \frac{Tr}{J} = \frac{(9000 \text{ N·m})\left(\dfrac{10 \text{ cm}}{(2)\left(100 \ \frac{\text{cm}}{\text{m}}\right)} \right)}{6.71 \times 10^{-6} \text{ m}^4}$$
$$= 67.05 \times 10^6 \text{ Pa} \quad (67 \text{ MPa})$$

The answer is (C).

14. Assume a thin-walled tank. Solve the equation for tangential (hoop) stress for the wall thickness. Although the inner radius is used by convention, the outer radius can be used.

$$t = \frac{pd}{2\sigma_t} = \frac{p(d_o - 2t)}{2\sigma_t} \approx \frac{pd_o}{2\sigma_t}$$
$$= \frac{(8 \text{ MPa})(25 \text{ cm})}{(2)(90 \text{ MPa})}$$
$$= 1.11 \text{ cm} \quad (1.1 \text{ cm})$$

Check the thin-wall assumption.

$$\frac{t}{r_i} = \frac{t}{\dfrac{d_o - 2t}{2}} = \frac{1.11 \text{ cm}}{\dfrac{25 \text{ cm} - (2)(1.11 \text{ cm})}{2}}$$
$$= 0.098 < 0.1 \quad [\text{thin wall}]$$

The answer is (C).

32 Beams

PRACTICE PROBLEMS

1. For the beam shown, what is most nearly the maximum compressive stress at section D-D, 1.5 m from the left end?

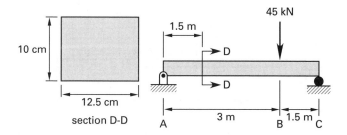

section D-D

(A) 63 MPa

(B) 110 MPa

(C) 230 MPa

(D) 330 MPa

2. Refer to the beam shown. The beam is fixed at one end. The beam has a mass of 46.7 kg/m. The modulus of elasticity of the beam is 200 GPa; the moment of inertia is 4680 cm^4.

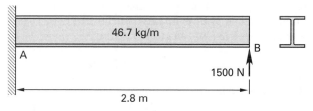

The upward force at B is 1500 N. What is most nearly the net deflection of the beam at a point 1.2 m from the fixed end?

(A) −0.32 mm (downward)

(B) −0.29 mm (downward)

(C) 0.12 mm (upward)

(D) 0.17 mm (upward)

3. Refer to the simply supported beam shown.

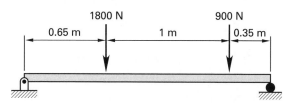

What is most nearly the maximum bending moment?

(A) 340 N·m

(B) 460 N·m

(C) 660 N·m

(D) 890 N·m

4. Refer to the cantilevered structural section shown. The beam is manufactured from steel with a modulus of elasticity of 210 GPa. The beam's cross-sectional area is 37.9 cm^2; its moment of inertia is 2880 cm^4. The beam has a mass of 45.9 kg/m. A 6000 N compressive force is applied at the top of the beam, at an angle of 30° from the horizontal. Neglect buckling.

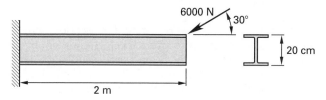

What is most nearly the maximum shear force in the beam?

(A) 3000 N

(B) 3900 N

(C) 5200 N

(D) 6100 N

5. For the cantilever steel rod shown, what is most nearly the force, F, necessary to deflect the rod a vertical distance of 7.5 mm?

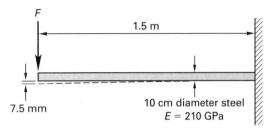

(A) 6900 N

(B) 8800 N

(C) 11 000 N

(D) 17 000 N

6. Refer to the simply supported beam shown.

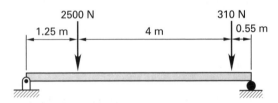

What is most nearly the maximum shear?

(A) 500 N

(B) 1000 N

(C) 1500 N

(D) 2000 N

7. Refer to the cantilevered structural section shown. The beam is manufactured from steel with a modulus of elasticity of 200 GPa. The beam's cross-sectional area is 74 cm^2; its moment of inertia is 8700 cm^4. The beam has a mass of 60 kg/m. A 2500 N compressive force is applied at the top of the beam, at an angle of 22° from horizontal. Neglect buckling.

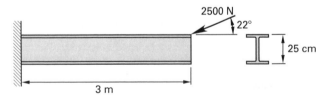

What is most nearly the approximate maximum bending moment in the beam?

(A) 5000 N·m

(B) 5200 N·m

(C) 5900 N·m

(D) 6100 N·m

8. A rectangular beam has a cross section of 5 cm wide × 10 cm deep and experiences a maximum shear of 2250 N. What is most nearly the maximum shear stress in the beam?

(A) 450 kPa

(B) 570 kPa

(C) 680 kPa

(D) 790 kPa

9. A simply supported beam supports a triangular distributed load as shown. The peak load at the right end of the beam is 5 N/m.

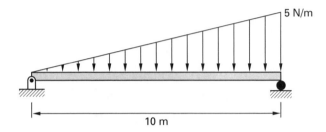

What is the approximate bending moment at a point 7 m from the left end of the beam?

(A) 15 N·m

(B) 17 N·m

(C) 28 N·m

(D) 30 N·m

10. Refer to the cantilevered structural section shown. The beam is manufactured from steel with a modulus of elasticity of 205 GPa. The beam's cross-sectional area is 86 cm^2; its moment of inertia is 24 400 cm^4. A 3700 N compressive force is applied at the top of the beam, at an angle of 40° from horizontal. A counterclockwise moment of 600 N·m is applied to the free end. Neglect beam self-weight, and neglect buckling.

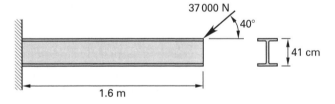

What is most nearly the deflection at the tip of the beam due to the external force alone (i.e., neglecting the beam's own mass)?

(A) 0.63 mm

(B) 0.82 mm

(C) 1.2 mm

(D) 2.5 mm

SOLUTIONS

1. Find the reaction at A.

$$\sum M_C = R_A(4.5 \text{ m}) - (45 \text{ kN})(1.5 \text{ m}) = 0$$
$$R_A = 15 \text{ kN}$$

The bending moment at section D-D is

$$M = (15 \text{ kN})(1.5 \text{ m}) = 22.5 \text{ kN·m}$$

The maximum compressive stress is at the top fiber of the beam section.

$$\sigma_{\max} = \frac{Mc}{I} = \frac{M\dfrac{h}{2}}{\dfrac{bh^3}{12}}$$
$$= \frac{(22.5 \text{ kN·m})\left(1000 \dfrac{\text{N}}{\text{kN}}\right)\left(\dfrac{0.10 \text{ m}}{2}\right)}{\dfrac{(0.125 \text{ m})(0.10 \text{ m})^3}{12}}$$
$$= 108 \times 10^6 \text{ Pa} \quad (110 \text{ MPa})$$

The answer is (B).

2. Use the principle of superposition to determine the deflection. The total deflection is the upward deflection due to the concentrated force less the downward deflection due to the weight of the beam.

Distance x is measured from the fixed end. The upward deflection due to the concentrated force is

$$v_{x,1} = \left(\frac{Px^2}{6EI}\right)(3L - x)$$
$$= \frac{(1500 \text{ N})(1.2 \text{ m})^2\left(100 \dfrac{\text{cm}}{\text{m}}\right)^4}{(6)(200 \text{ GPa})\left(10^9 \dfrac{\text{Pa}}{\text{GPa}}\right)(4680 \text{ cm}^4)}$$
$$\times \big((3)(2.8 \text{ m}) - 1.2 \text{ m}\big)$$
$$= 0.000277 \text{ m} \quad (0.28 \text{ mm}) \quad [\text{upward}]$$

The downward deflection is due to the beam's own mass. Distance x is measured from the fixed end. The load per unit length is

$$w = mg = \left(46.7 \frac{\text{kg}}{\text{m}}\right)\left(9.81 \frac{\text{m}}{\text{s}^2}\right) = 458 \text{ N/m}$$

The downward deflection is

$$v_{x,2} = \left(\frac{-wx^2}{24EI}\right)(x^2 - 4Lx + 6L^2)$$
$$= \left(\frac{-\left(458 \dfrac{\text{N}}{\text{m}}\right)(1.2 \text{ m})^2\left(100 \dfrac{\text{cm}}{\text{m}}\right)^4}{(24)(200 \text{ GPa})\left(10^9 \dfrac{\text{Pa}}{\text{GPa}}\right)(4680 \text{ cm}^4)}\right)$$
$$\times \left(\begin{array}{c}(1.2 \text{ m})^2 - (4)(2.8 \text{ m})(1.2 \text{ m}) \\ + (6)(2.8 \text{ m})^2\end{array}\right)$$
$$= -0.000103 \text{ m} \quad (-0.10 \text{ mm}) \quad [\text{downward}]$$

The net deflection is

$$v = v_{x,1} + v_{x,2} = 0.28 \text{ mm} + (-0.10 \text{ mm})$$
$$= 0.17 \text{ mm} \quad [\text{upward}]$$

The answer is (D).

3. Determine the reactions by taking moments about each end.

$$\sum M_B = -R_A(0.65 \text{ m} + 1 \text{ m} + 0.35 \text{ m})$$
$$+ (1800 \text{ N})(1 \text{ m} + 0.35 \text{ m})$$
$$+ (900 \text{ N})(0.35 \text{ m}) = 0$$
$$R_A = 1372.5 \text{ N}$$
$$\sum F_y = R_B + 1372.5 \text{ N} - 1800 \text{ N} - 900 \text{ N} = 0$$
$$R_B = 1327.5 \text{ N}$$

Draw the shear and moment diagrams.'

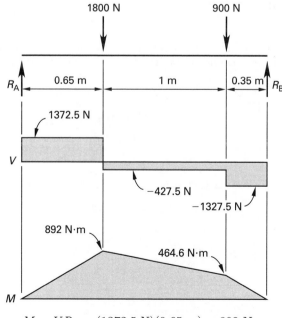

$$M = VR_A = (1372.5 \text{ N})(0.65 \text{ m}) = 892 \text{ N·m}$$

The maximum moment occurs 0.65 m from the left end (where V goes through zero) of the beam and is equal to 892 N·m (890 N·m).

The answer is (D).

4. The maximum vertical shear in the beam will occur at the fixed end.

$$
\begin{aligned}
V &= wL + F_y \\
&= mgL + F_y \\
&= \left(45.9\ \frac{\text{kg}}{\text{m}}\right)\left(9.81\ \frac{\text{m}}{\text{s}^2}\right)(2\text{ m}) + (6000\text{ N})(\sin 30°) \\
&= 3900\text{ N}
\end{aligned}
$$

The answer is (B).

5. For a cantilever beam loaded at its tip, with $x = L$,

$$
v_{\max} = \frac{-PL^3}{3EI}
$$

$$
\begin{aligned}
P &= \frac{-3EIv_{\max}}{L^3} \\
&= \frac{-(3)(210\text{ GPa})\left(10^9\ \dfrac{\text{Pa}}{\text{GPa}}\right)}{(1.5\text{ m})^3\left(1000\ \dfrac{\text{mm}}{\text{m}}\right)} \\
&\quad \times \left(\frac{\pi}{4}\right)(0.05\text{ m})^4(7.5\text{ mm}) \\
&= -6872\text{ N} \quad (6900\text{ N}) \quad [\text{downward}]
\end{aligned}
$$

The answer is (A).

6. Determine the reactions by taking the moments about end B and by taking the sum of the forces.

$$
\begin{aligned}
\sum M_B &= -R_A(1.25\text{ m} + 4\text{ m} + 0.55\text{ m}) \\
&\quad + (2500\text{ N})(4\text{ m} + 0.55\text{ m}) \\
&\quad + (310\text{ N})(0.55\text{ m}) \\
&= 0 \\
R_A &= 1990.6\text{ N} \\
\sum F_y &= R_B + 1990.6\text{ N} - 2500\text{ N} - 310\text{ N} \\
&= 0 \\
R_B &= 819.4\text{ N}
\end{aligned}
$$

Draw the shear diagram.

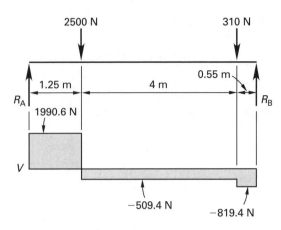

From the shear diagram, the maximum shear is 1990.6 N (2000 N).

The answer is (D).

7. The maximum bending moment will occur at the fixed end of the beam. The moment will be affected by the distributed load and the external force. Since the force does not act through the centroid of the beam (i.e., the force is eccentric), both the vertical and the horizontal components of the external force must be included.

The moment due to the beam's own mass is

$$
\begin{aligned}
M_1 &= \tfrac{1}{2}wL^2 = \tfrac{1}{2}mgL^2 = \left(\tfrac{1}{2}\right)\left(60\ \frac{\text{kg}}{\text{m}}\right)\left(9.81\ \frac{\text{m}}{\text{s}^2}\right)(3\text{ m})^2 \\
&= 2648.7\text{ N·m}
\end{aligned}
$$

The moment due to the vertical component of the external force is

$$
\begin{aligned}
M_2 &= F_yL = (2500\text{ N})(\sin 22°)(3\text{ m}) \\
&= 2809.5\text{ N·m}
\end{aligned}
$$

The force is not applied through the beam's centroid. The horizontal component of the force causes the beam to bend upward, while the other forces bend the beam downward. The moment due to the eccentricity is

$$
\begin{aligned}
M_3 &= -F_xe = -(2500\text{ N})(\cos 22°)\left(\frac{25\text{ cm}}{(2)\left(100\ \dfrac{\text{cm}}{\text{m}}\right)}\right) \\
&= -289.7\text{ N·m}
\end{aligned}
$$

The total moment is

$$
\begin{aligned}
M &= M_1 + M_2 + M_3 \\
&= 2648.7\text{ N·m} + 2809.5\text{ N·m} + (-298.7\text{ N·m}) \\
&= 5168.5\text{ N·m} \quad (5200\text{ N·m})
\end{aligned}
$$

The answer is (B).

8. The maximum shear stress is

$$\tau_{max} = \frac{3V}{2A} = \frac{(3)(2250 \text{ N})\left(100 \frac{\text{cm}}{\text{m}}\right)^2}{(2)(5 \text{ cm})(10 \text{ cm})}$$

$$= 675 \times 10^3 \text{ Pa} \quad (680 \text{ kPa})$$

The answer is (C).

9. The total force from the distributed load is

$$\left(\tfrac{1}{2}\right)(10 \text{ m})\left(5 \frac{\text{N}}{\text{m}}\right) = 25 \text{ N}$$

This force can be assumed to act at two-thirds of the beam length from the left end, or one-third of the beam length from the right end.

Sum the moments around the right end to find the left reaction.

$$\sum M_{\text{right end}} = (25 \text{ N})\left(\frac{10 \text{ m}}{3}\right) - R_{\text{left}}(10 \text{ m}) = 0$$

$$R_{\text{left}} = 8.33 \text{ N}$$

The load increases linearly to 5 N/m at 10 m. At 7 m, the loading is $(0.7)(5 \text{ N/m})$. The total distributed force over the first 7 m of the beam is

$$\left(\tfrac{1}{2}\right)(7 \text{ m})\left((0.7)\left(5 \frac{\text{N}}{\text{m}}\right)\right) = 12.25 \text{ N}$$

Sum moments from the point of interest (7 m from the left end) to either end. The calculation is easier from the left end.

$$\sum M = (12.25 \text{ N})\left(\frac{7 \text{ m}}{3}\right) - (8.33 \text{ N})(7 \text{ m})$$

$$= -29.73 \text{ N·m} \quad (30 \text{ N·m})$$

The answer is (D).

10. With $x = L$, the deflection due to the vertical component of the force is

$$v_1 = \frac{-PL^3}{3EI} = \frac{-(37\,000 \text{ N})(\sin 40°)(1.6 \text{ m})^3\left(100 \frac{\text{cm}}{\text{m}}\right)^4}{(3)(205 \text{ GPa})\left(10^9 \frac{\text{Pa}}{\text{GPa}}\right)(24\,400 \text{ cm}^4)}$$

$$= -0.000649 \text{ m} \quad (-0.649 \text{ mm}) \quad [\text{downward}]$$

The eccentric application of the force causes an upward deflection. The deflection due to the end moment is

$$v_2 = \frac{M_0 L^2}{2EI} = \frac{(600 \text{ N·m})(1.6 \text{ m})^2\left(100 \frac{\text{cm}}{\text{m}}\right)^4}{(2)(205 \text{ GPa})\left(10^9 \frac{\text{Pa}}{\text{GPa}}\right)(24\,400 \text{ cm}^4)}$$

$$= 0.0000154 \text{ m} \quad (0.0154 \text{ mm}) \quad [\text{upward}]$$

The total deflection due to the external force alone is

$$v = v_1 + v_2 = -0.649 \text{ mm} + 0.0154 \text{ mm}$$

$$= -0.634 \text{ mm} \quad (0.63 \text{ mm}) \quad [\text{downward}]$$

The answer is (A).

33 Columns

PRACTICE PROBLEMS

1. A steel column with a cross section of 12 cm × 16 cm is 4 m in height and fixed at its base. The column is pinned against translation in its weak direction at the top but is unbraced in its strong direction. The column's modulus of elasticity is 2.1×10^5 MPa.

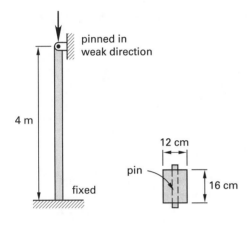

What is most nearly the maximum theoretical vertical load the column can support without buckling?

(A) 1.3 MN

(B) 5.2 MN

(C) 6.1 MN

(D) 11 MN

2. A 10 cm × 10 cm square column supports a compressive force of 9000 N. The load is applied with an eccentricity of 2.5 cm along one of the lines of symmetry. What is most nearly the maximum tensile stress in the column?

(A) 450 kPa

(B) 900 kPa

(C) 1400 kPa

(D) 2300 kPa

3. A square column with a solid cross section is placed in a building to support a load of 5 MN. The maximum allowable stress in the column is 350 MPa. The column reacts linearly to all loads. If the contractor is permitted to load the column anywhere in the central one-fifth of the column's cross section, what are most nearly the smallest possible dimensions of the column?

(A) 12 cm × 12 cm

(B) 14 cm × 14 cm

(C) 16 cm × 16 cm

(D) 18 cm × 18 cm

4. What is most nearly the maximum resultant normal stress at A for the cantilever beam shown?

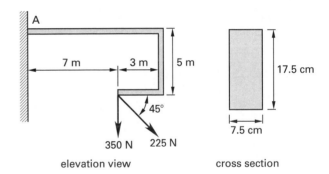

elevation view   cross section

(A) 7.2 MPa

(B) 9.4 MPa

(C) 9.8 MPa

(D) 9.9 MPa

5. A rectangular steel bar 37.5 mm wide and 50 mm thick is pinned at each end and subjected to axial compression. The bar has a length of 1.75 m. The modulus of elasticity is 200 GPa. What is most nearly the critical buckling load?

(A) 60 kN

(B) 93 kN

(C) 110 kN

(D) 140 kN

6. What is most nearly the Euler buckling load for a 10 m long steel column with unrestrained ends and with the given properties and cross section?

$$I_{x'x'} = 3.70 \times 10^6 \text{ mm}^4$$

$$E = 200 \text{ GPa}$$

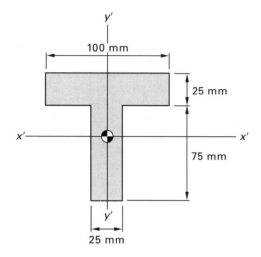

(A) 15 kN

(B) 24 kN

(C) 43 kN

(D) 73 kN

SOLUTIONS

1. Since the column is fixed at one end and pinned at the other, the theoretical end-restraint coefficient, K, is 0.7. The effective length for buckling in the weak direction is

$$K\ell = (0.7)(4 \text{ m}) = 2.8 \text{ m}$$

The moment of inertia for buckling in the weak direction is

$$I = \frac{bh^3}{12} = \frac{(16 \text{ cm})(12 \text{ cm})^3}{(12)\left(100 \ \frac{\text{cm}}{\text{m}}\right)^4}$$

$$= 2.3 \times 10^{-5} \text{ m}^4$$

Calculate the critical buckling force from Euler's formula.

$$P_{cr} = \frac{\pi^2 EI}{(K\ell)^2}$$

$$= \frac{\pi^2 (2.1 \times 10^5 \text{ MPa})\left(10^6 \ \frac{\text{Pa}}{\text{MPa}}\right)(2.3 \times 10^{-5} \text{ m}^4)}{(2.8 \text{ m})^2}$$

$$= 6.09 \times 10^6 \text{ N} \quad (6.1 \text{ MN})$$

Check the buckling force in the strong direction. The column is not braced in that direction, so for a column fixed at one end and free at the other, $K = 2$.

$$K\ell = (2)(4 \text{ m}) = 8 \text{ m}$$

The moment of inertia for buckling in the strong direction is

$$I = \frac{bh^3}{12} = \frac{(12 \text{ cm})(16 \text{ cm})^3}{(12)\left(100 \ \frac{\text{cm}}{\text{m}}\right)^4} = 4.1 \times 10^{-5} \text{ m}^4$$

Calculate the critical buckling force from Euler's formula.

$$P_{cr} = \frac{\pi^2 EI}{(K\ell)^2}$$

$$= \frac{\pi^2 (2.1 \times 10^5 \text{ MPa})\left(10^6 \ \frac{\text{Pa}}{\text{MPa}}\right)(4.1 \times 10^{-5} \text{ m}^4)}{(8 \text{ m})^2}$$

$$= 1.3 \times 10^6 \text{ N} \quad (1.3 \text{ MN})$$

This is less than for buckling in the weak direction. This force controls.

The answer is (A).

2. The cross-sectional area of a square column is

$$A = b^2 = \left(\frac{10 \text{ cm}}{100 \frac{\text{cm}}{\text{m}}}\right)^2 = 0.01 \text{ m}^2$$

The moment of inertia of the square cross section is

$$I = \frac{b^4}{12} = \frac{(10 \text{ cm})^4}{(12)\left(100 \frac{\text{cm}}{\text{m}}\right)^4} = 8.33 \times 10^{-6} \text{ m}^4$$

The distance from the neutral axis to the extreme fibers is

$$c = \frac{b}{2} = \frac{10 \text{ cm}}{(2)\left(100 \frac{\text{cm}}{\text{m}}\right)} = 0.05 \text{ m}$$

The stress is

$$\sigma = \frac{F}{A} \pm \frac{Fec}{I}$$
$$= \frac{-9000 \text{ N}}{0.01 \text{ m}^2} \pm \frac{(-9000 \text{ N})(2.5 \text{ cm})(0.05 \text{ m})}{(8.33 \times 10^{-6} \text{ m}^4)\left(100 \frac{\text{cm}}{\text{m}}\right)}$$
$$= -9 \times 10^5 \text{ Pa} \pm 1.35 \times 10^6 \text{ Pa}$$
$$(-900 \text{ kPa} \pm 1350 \text{ kPa})$$

The first term is due to the compressive column load and is compressive. (Compressive forces and stresses are usually given a negative sign.) The second term is due to the eccentricity. The second term increases the compressive stress at the inner face. It counteracts the compressive stress at the outer face.

The maximum tensile stress is

$$\sigma_{t,\text{max}} = -900 \text{ kPa} + 1350 \text{ kPa} = 450 \text{ kPa}$$

The answer is (A).

3. The middle one-fifth of the column is a square with dimensions of $b/5 \times b/5$ ($0.2b \times 0.2b$).

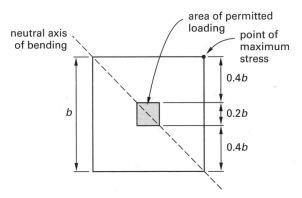

The maximum stress will be induced when the middle one-fifth square is loaded at one of its corners.

The cross-sectional area is

$$A = b^2$$

The moment of inertia of the square cross section is

$$I = \frac{b^4}{12}$$

The distance from the neutral axis to the extreme fibers is

$$c = \frac{b}{2}$$

The maximum eccentricity is

$$e = 0.1b$$

The stress at the extreme corner is

$$\sigma = \frac{F}{A} \pm \frac{Fe_x c_x}{I_x} + \frac{Fe_y c_y}{I_y}$$
$$= F\left(\frac{1}{b^2} \pm \frac{(2)(0.1b)\left(\frac{b}{2}\right)}{\frac{b^4}{12}}\right)$$
$$= F\left(\frac{1}{b^2} \pm \frac{1.2}{b^2}\right)$$
$$= \frac{2.2F}{b^2}$$
$$b = \sqrt{\frac{2.2F}{\sigma}}$$
$$= \sqrt{\frac{(2.2)(5 \text{ MN})\left(10^6 \frac{\text{N}}{\text{MN}}\right)}{(350 \text{ MPa})\left(10^6 \frac{\text{Pa}}{\text{MPa}}\right)}}$$
$$= 0.177 \text{ m} \quad (18 \text{ cm})$$

The answer is (D).

4. The beam experiences both axial tension and bending stresses, so it should be analyzed as a beam-column.

$$\sum M_A = (350 \text{ N})(7 \text{ m}) + (225 \text{ N})(\sin 45°)(7 \text{ m})$$
$$- (225 \text{ N})(\cos 45°)(5 \text{ m})$$
$$= 2768 \text{ N·m}$$

The stress is

$$\sigma_{max} = \frac{P}{A} + \frac{Mc}{I} = \frac{P}{bh} + \frac{M\left(\frac{b}{2}\right)}{\frac{bh^3}{12}}$$

$$= \frac{(225 \text{ N})(\cos 45°)\left(100 \; \frac{\text{cm}}{\text{m}}\right)^2}{(7.5 \text{ cm})(17.5 \text{ cm})}$$

$$+ \frac{(2768 \text{ N·m})\left(\dfrac{17.5 \text{ cm}}{(2)\left(100 \; \frac{\text{cm}}{\text{m}}\right)}\right)}{\dfrac{(7.5 \text{ cm})(17.5 \text{ cm})^3}{(12)\left(100 \; \frac{\text{cm}}{\text{m}}\right)^4}}$$

$$= 7.24 \times 10^6 \text{ Pa} \quad (7.2 \text{ MPa})$$

The answer is (A).

5. Use Euler's formula. $K = 1$ since both ends are pinned. Use the moment of inertia for the weak direction.

$$P_{cr} = \frac{\pi^2 EI}{(K\ell)^2} = \frac{\pi^2 E\left(\frac{bh^3}{12}\right)}{(K\ell)^2}$$

$$= \frac{\pi^2 (200 \text{ GPa})\left(10^9 \; \frac{\text{Pa}}{\text{GPa}}\right)\left(\dfrac{(50 \text{ mm})(37.5 \text{ mm})^3}{(12)\left(1000 \; \frac{\text{mm}}{\text{m}}\right)^4}\right)}{((1)(1.75 \text{ m}))^2}$$

$$= 141\,624 \text{ N} \quad (140 \text{ kN})$$

The answer is (D).

6. $x'x'$ and $y'y'$ are centroidal axes. $I_{y'y'}$ is computed from the equation $I = bh^3/12$ about the centroidal axis of a rectangle. For this cross section, $b_1 = 25$ mm, $h_1 = 100$ mm, $b_2 = 75$ mm, and $h_2 = 25$ mm.

$$I_{y'y'} = \frac{b_1 h_1^3}{12} + \frac{b_2 h_2^3}{12}$$

$$= \frac{(25 \text{ mm})(100 \text{ mm})^3}{(12)\left(1000 \; \frac{\text{mm}}{\text{m}}\right)^4} + \frac{(75 \text{ mm})(25 \text{ mm})^3}{(12)\left(1000 \; \frac{\text{mm}}{\text{m}}\right)^4}$$

$$= 2.18 \times 10^{-6} \text{ m}^4$$

Find the Euler buckling load, P_{cr}. The smallest moment of inertia (corresponding to the least radius of gyration) should be used. $I_{y'y'}$ is less than $I_{x'x'}$.

$$P_{cr} = \frac{\pi^2 EI}{(K\ell)^2}$$

$$= \frac{\pi^2 (200 \text{ GPa})\left(10^6 \; \frac{\text{kPa}}{\text{GPa}}\right)(2.18 \times 10^{-6} \text{ m}^4)}{((1)(10 \text{ m}))^2}$$

$$= 43 \text{ kN}$$

The answer is (C).

34 Material Properties and Testing

PRACTICE PROBLEMS

1. A stress-strain diagram is shown.

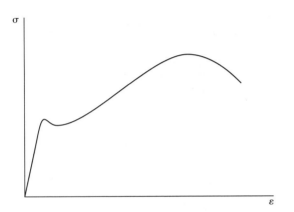

What test is represented by the illustration?

(A) resilience test

(B) rotating beam test

(C) ductility test

(D) tensile test

2. A 0.4 m long steel rod has a diameter of 0.05 m and a modulus of elasticity of 20×10^4 MPa. The rod supports a 10 000 N compressive load. What is most nearly the decrease in the steel rod's length?

(A) 1.3×10^{-6} m

(B) 2.5×10^{-6} m

(C) 5.1×10^{-6} m

(D) 1.0×10^{-5} m

3. What is the term for the ratio of stress to strain below the proportional limit?

(A) modulus of rigidity

(B) Hooke's constant

(C) Poisson's ratio

(D) Young's modulus

4. What do impact tests determine?

(A) hardness

(B) yield strength

(C) toughness

(D) creep strength

5. The density of a particular metal is 2750 kg/m^3. The modulus of elasticity for this metal is 210 GPa. A circular bar of this metal 3.5 m long and 160 cm^2 in cross-sectional area is suspended vertically from one end. What is most nearly the elongation of the bar due to its own mass?

(A) 0.00055 mm

(B) 0.00079 mm

(C) 0.0016 mm

(D) 0.0024 mm

6. A stress-strain diagram is shown.

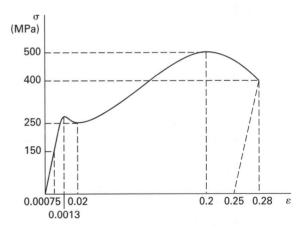

What is most nearly the percent elongation at failure?

(A) 14%

(B) 19%

(C) 25%

(D) 28%

Materials

7. What does the value of 40 MPa in the illustration shown represent?

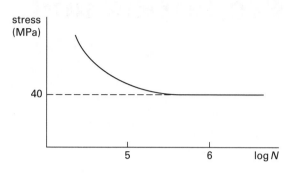

I.   fatigue limit

II.   endurance limit

III.   proportional limit

IV.   yield stress

(A) I only

(B) I and II

(C) II and IV

(D) I, II, and IV

8. If δ is deformation, and L is the original length of the specimen, what is the definition of normal strain, ε?

(A) $\varepsilon = \dfrac{L+\delta}{L}$

(B) $\varepsilon = \dfrac{L+\delta}{\delta}$

(C) $\varepsilon = \dfrac{\delta}{L+\delta}$

(D) $\varepsilon = \dfrac{\delta}{L}$

9. Which of the following statements regarding the ductile-to-brittle transition temperature is true?

I.   It is important for structures used in cold environments.

II.   It is the point at which the size of the shear lip or tearing rim goes to zero.

III.   It is the temperature at which 20 J of energy causes failure in a Charpy V-notch specimen of standard dimensions.

(A) I only

(B) I and II

(C) I and III

(D) II and III

10. A stress-strain diagram is shown.

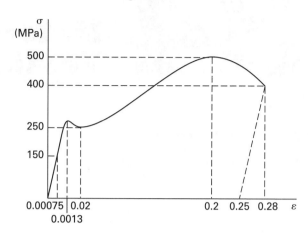

What is most nearly the modulus of elasticity of the material?

(A) 20 GPa

(B) 80 GPa

(C) 100 GPa

(D) 200 GPa

SOLUTIONS

1. The illustration shows results from a tensile test. Both resilience and ductility may be calculated from the results, but the test is not known by those names. The rotating beam is a cyclic test and does not yield a monotonic stress-strain curve.

The answer is (D).

2. The area of the steel rod is

$$A_0 = \frac{\pi d^2}{4} = \frac{\pi (0.05 \text{ m})^2}{4} = 1.96 \times 10^{-3} \text{ m}^2$$

The decrease in the rod's length is

$$\Delta L = \frac{FL_0}{A_0 E}$$

$$= \frac{(10\,000 \text{ N})(0.4 \text{ m})}{(1.96 \times 10^{-3} \text{ m}^2)(20 \times 10^4 \text{ MPa})\left(10^6 \frac{\text{Pa}}{\text{MPa}}\right)}$$

$$= 1.019 \times 10^{-5} \text{ m} \quad (1.0 \times 10^{-5} \text{ m})$$

The answer is (D).

3. Young's modulus is defined by Hooke's law.

$$\sigma = E\varepsilon$$

E is Young's modulus, or the modulus of elasticity, equal to the stress divided by strain within the proportional region of the stress-strain curve.

The answer is (D).

4. An impact test measures the energy needed to fracture the test sample. This is a toughness parameter.

The answer is (C).

5. The mass of the bar is

$$m = \rho V = \rho A L$$

$$= \left(2750 \frac{\text{kg}}{\text{m}^3}\right) \left(\frac{160 \text{ cm}^2}{\left(100 \frac{\text{cm}}{\text{m}}\right)^2}\right)(3.5 \text{ m})$$

$$= 154 \text{ kg}$$

The total gravitational force is experienced by the metal at the suspension point. Farther down the rod, however, there is less volume contributing to the force, and the stress is reduced. The average force on the metal in the bar is half of the maximum value.

$$F_{\text{ave}} = \tfrac{1}{2} F_{\text{max}} = \tfrac{1}{2} mg$$

$$= \left(\tfrac{1}{2}\right)(154 \text{ kg})\left(9.81 \frac{\text{m}}{\text{s}^2}\right)$$

$$= 755 \text{ N}$$

The elongation is

$$\varepsilon = \frac{\Delta L}{L_0}$$

$$\Delta L = \varepsilon L_0 = \frac{\sigma}{E} L_0 = \frac{F}{AE} L_0$$

$$= \left(\frac{755 \text{ N}}{(160 \text{ cm}^2)(210 \times 10^9 \text{ Pa})}\right)(3.5 \text{ m})\left(100 \frac{\text{cm}}{\text{m}}\right)^2$$

$$= 7.868 \times 10^{-7} \text{ m} \quad (0.00079 \text{ mm})$$

The answer is (B).

6. The strain at failure used in the equation is found by extending a line from the failure point to the strain axis, parallel to the linear portion of the curve. The percent elongation is an indicator of the ductility of a material, but it is not the same as the ductility. The percent elongation is

$$\text{percent elongation} = \varepsilon_f \times 100\%$$

$$= 0.25 \times 100\%$$

$$= 25\%$$

The answer is (C).

7. The illustration shows results of an endurance (or fatigue) test. The value of 40 MPa is called the endurance stress, endurance limit, or fatigue limit, and is equal to the maximum stress that can be repeated indefinitely without causing the specimen to fail.

The answer is (B).

8. Strain is defined as elongation, δ, per unit length, L.

The answer is (D).

9. Option II is the only one that is false. A test piece that breaks at 20 J of energy usually has a small shear lip.

The answer is (C).

10. The modulus of elasticity (Young's modulus) is the slope of the stress-strain line in the proportional region.

$$\sigma = E\varepsilon$$

$$E = \frac{\sigma}{\varepsilon} = \frac{150 \text{ MPa}}{0.00075}$$

$$= 200\,000 \text{ MPa} \quad (20 \times 10^4 \text{ MPa})$$

The answer is (D).

Materials

35 Engineering Materials

PRACTICE PROBLEMS

1. Refer to the phase diagram shown.

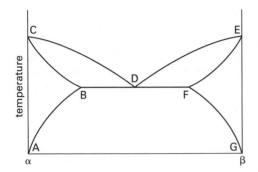

The region enclosed by points DEF can be described as a

- (A) mixture of solid β component and liquid α component
- (B) mixture of solid β and liquid β component
- (C) peritectic composition
- (D) mixture of solid β component and a molten mixture of α and β components

2. Which of the following figures is a cooling curve of a pure metal?

(A)

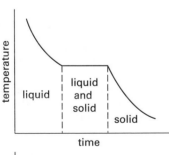

(B)

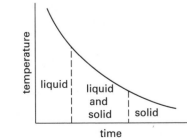

(C)

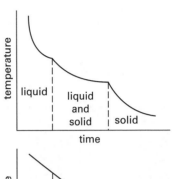

(D)

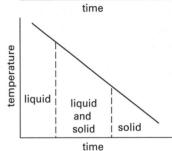

3. A composite material consists of 20 kg of material A, 10 kg of material B, and 5 kg of material C. The densities of materials A, B, and C are 2 g/cm^3, 3 g/cm^3, and 4 g/cm^3, respectively. What is most nearly the density of the composite material?

- (A) 2.1 g/cm^3
- (B) 2.4 g/cm^3
- (C) 2.7 g/cm^3
- (D) 3.3 g/cm^3

4. Which of the following characteristics describes martensite?

I. high ductility

II. formed by quenching austenite

III. high hardness

- (A) I only
- (B) I and II
- (C) I and III
- (D) II and III

5. A mixture of ice and water is held at a constant temperature of 0°C. How many degrees of freedom does the mixture have?

(A) −1

(B) 0

(C) 1

(D) 2

6. Given the electrochemical cell shown, what is the reaction at the anode?

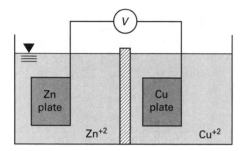

(A) $Cu \rightarrow Cu^{2+} + 2e^-$

(B) $Cu^{2+} + 2e^- \rightarrow Cu$

(C) $Zn \rightarrow Zn^{2+} + 2e^-$

(D) $Zn^{2+} + 2e^- \rightarrow Zn$

7. Refer to the phase diagram shown.

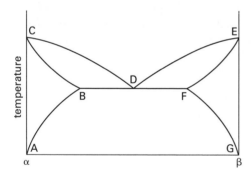

Which line(s) is/are the liquidus line(s)?

(A) CBDFG

(B) CDE

(C) CBFE

(D) ABC and EFG

8. What is the hardest form of steel?

(A) pearlite

(B) ferrite

(C) bainite

(D) martensite

9. Which of the following processes can increase the deformation resistance of steel?

I.   tempering

II.   hot working

III.   adding alloying elements

IV.   hardening

(A) I and II

(B) I and IV

(C) II and III

(D) III and IV

10. Corrosion of iron can be inhibited with a more electropositive coating, while a less electropositive coating tends to accelerate corrosion. Which of the following coatings will contribute to corrosion of iron products?

(A) zinc

(B) gold

(C) aluminum

(D) magnesium

11. Refer to the phase diagram shown.

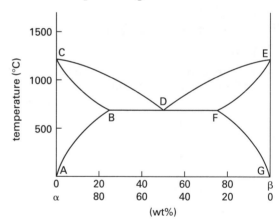

Approximately how much solid (as a percentage by weight) exists when the mixture is 30% α and 70% β and the temperature is 800°C?

(A) 0%

(B) 19%

(C) 30%

(D) 50%

SOLUTIONS

1. The region describes a mixture of solid β component and a liquid of components α and β.

The answer is (D).

2. The solidification of a molten metal is no different than the solidification of water into ice. During the phase change, the temperature remains constant as the heat of fusion is removed. The temperature remains constant during the phase change.

The answer is (A).

3. Calculate the volume, V, of each material.

$$V_A = \frac{m_A}{\rho_A} = \frac{(20 \text{ kg})\left(1000 \frac{\text{g}}{\text{kg}}\right)}{2 \frac{\text{g}}{\text{cm}^3}}$$

$$= 10\,000 \text{ cm}^3$$

$$V_B = \frac{m_B}{\rho_B} = \frac{(10 \text{ kg})\left(1000 \frac{\text{g}}{\text{kg}}\right)}{3 \frac{\text{g}}{\text{cm}^3}}$$

$$= 3333 \text{ cm}^3$$

$$V_C = \frac{m_C}{\rho_C} = \frac{(5 \text{ kg})\left(1000 \frac{\text{g}}{\text{kg}}\right)}{4 \frac{\text{g}}{\text{cm}^3}}$$

$$= 1250 \text{ cm}^3$$

The total volume is

$$V_{\text{tot}} = V_A + V_B + V_C$$

$$= 10\,000 \text{ cm}^3 + 3333 \text{ cm}^3 + 1250 \text{ cm}^3$$

$$= 14\,583 \text{ cm}^3$$

The density of the composite material is

$$\rho_c = \sum f_i \rho_i = f_A \rho_A + f_B \rho_B + f_C \rho_C$$

$$= \frac{V_A \rho_A + V_B \rho_B + V_C \rho_C}{V_{\text{tot}}}$$

$$= \frac{\begin{array}{l}(10\,000 \text{ cm}^3)\left(2 \frac{\text{g}}{\text{cm}^3}\right) + (3333 \text{ cm}^3)\left(3 \frac{\text{g}}{\text{cm}^3}\right) \\ + (1250 \text{ cm}^3)\left(4 \frac{\text{g}}{\text{cm}^3}\right)\end{array}}{14\,583 \text{ cm}^3}$$

$$= 2.4 \text{ g/cm}^3$$

The answer is (B).

4. Martensite is a hard, strong, and brittle material formed by rapid cooling of austenite.

The answer is (D).

5. Since solid and liquid phases are present simultaneously, the number of phases, P, is 2. Only water is involved, so the number of compounds, C, is 1.

Gibbs' phase rule is applicable when both temperature and pressure can be varied. When the temperature is held constant, Gibbs' phase rule is

$$P + F = C + 1\Big|_{\text{constant temperature}}$$

$$F = C + 1 - P$$

$$= 1 + 1 - 2$$

$$= 0$$

The answer is (B).

6. Zinc has a higher potential and will act as the anode. By definition, the anode is where electrons are lost. The reaction at the anode of the electrochemical cell is $Zn \rightarrow Zn^{2+} + 2e^-$.

The answer is (C).

7. The liquidus line divides the diagram into two regions. Above the liquidus line, the alloy is purely liquid, while below the liquidus line, the alloy may exist as solid phase or as a mixture of solid and liquid phases. The liquidus line is CDE.

The answer is (B).

8. Hard steel is obtained by rapid quenching. Martensite has a high hardness since it is rapidly quenched. Though martensite is hard, it has low ductility.

The answer is (D).

9. Surface hardening processes will increase the deformation resistance of steel. Some alloying metals will also increase steel hardness. Tempering and hot working increase the ductility (deformation capability) of steel.

The answer is (D).

10. Zinc, aluminum, and magnesium are all more electropositive (anodic) than iron and will corrode sacrificially to protect it. Gold is more cathodic and will be protected at the expense of the iron.

The answer is (B).

11. Use the phase diagram to find the fraction of solid.

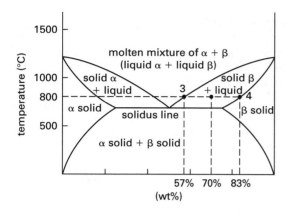

$$\text{wt\% fraction solid} = \frac{x - x_3}{x_4 - x_3} \times 100\%$$

$$= \frac{70\% - 57\%}{83\% - 57\%} \times 100\%$$

$$= 50\%$$

The answer is (D).

36 Structural Design: Materials and Basic Concepts

PRACTICE PROBLEMS

1. Which statement about the modulus of elasticity, E, is true?

(A) It is the same as the rupture modulus.

(B) It is the slope of the stress-strain curve in the linearly elastic region.

(C) It is the ratio of stress to volumetric strain.

(D) Its value depends only on the temperature of the material.

2. What modulus of elasticity is predicted by ACI 318 for normal weight concrete with a compressive strength of 3000 lbf/in^2?

(A) 0.26×10^6 psi

(B) 1.9×10^6 psi

(C) 2.8×10^6 psi

(D) 3.2×10^6 psi

3. Which of the following criteria must be met in order for the compressive strength of concrete to be satisfied?

I. No single strength test falls below the specified compressive strength, f'_c, by more than $0.10f'_c$.

II. No single strength test falls below the specified compressive strength, f'_c, by more than $0.20f'_c$.

III. The average of every three consecutive strength tests equals or exceeds the specified compressive strength.

IV. The average of every three consecutive strength tests must not equal or exceed the specified compressive strength.

(A) I and III

(B) I and IV

(C) II and III

(D) II and IV

4. A waste product of coal-burning power-generation stations, fly ash, is the most common pozzolamic additive. Which of the following statements are true about fly ash?

I. Fly ash reacts with calcium hydroxide to increase binding.

II. Fly ash reacts with calcium silicate to form a binder.

III. Fly ash acts as a microfiller between cement particles, increasing strength and durability while reducing permeability.

IV. When used as a replacement for less than 45% of the portland cement, fly ash meeting ASTM C618 enhances resistance to scaling from road-deicing chemicals.

(A) I and II

(B) III and IV

(C) I, III, and IV

(D) II, III, and IV

5. Which category of steel contains 0.15–0.29% carbon?

(A) low-carbon

(B) mild-carbon

(C) medium-carbon

(D) high-carbon

6. Which property of steel allows it to undergo large inelastic deformations without fracture?

(A) yield

(B) elasticity

(C) ductility

(D) toughness

7. A simply supported reinforced concrete beam 12 in wide and 28 in deep spans 20 ft. The beam is subjected to a uniform service dead load equal to 2.0 kips/ft (exclusive of beam weight) and to a uniform service live load of 2.4 kips/ft. The factored uniform load is most nearly

(A) 6.7 kips/ft

(B) 7.4 kips/ft

(C) 8.0 kips/ft

(D) 9.2 kips/ft

8. A simply supported beam 12 in wide, 24 in deep, and 30 ft long is subjected to a dead load of 1.2 kips/ft and a live load of 0.8 kip/ft in addition to its own dead weight. Most nearly, what moment should be used to determine the steel reinforcement at the center of the beam?

(A) 12 ft-kips

(B) 350 ft-kips

(C) 380 ft-kips

(D) 420 ft-kips

SOLUTIONS

1. The modulus of elasticity is the slope of the stress-strain diagram in the linearly elastic region.

The answer is (B).

2. The modulus of elasticity is

$$
\begin{aligned}
E_c &= 33 w_c^{1.5} \sqrt{f_c'} \\
&= (33)\left(145 \ \frac{\text{lbf}}{\text{in}^3}\right)^{1.5} \sqrt{3000 \ \frac{\text{lbf}}{\text{in}^2}} \\
&= 3.16 \times 10^6 \ \text{psi} \quad (3.2 \times 10^6 \ \text{psi})
\end{aligned}
$$

The answer is (D).

3. The compressive strength of concrete is considered satisfactory if two criteria are met: (a) no single strength test falls below the specified compressive strength, f_c', by more than $0.10 f_c'$, and (b) the average of every three consecutive strength tests equals or exceeds the specified compressive strength.

The answer is (A).

4. Fly ash reacts with calcium hydroxide to increase binding, acts as a microfiller between cement particles to increase strength and durability while decreasing permeability, and enhances resistance to scaling from road-deicing chemicals when it meets ASTM C618 and is used as a replacement for less than 45% of the portland cement.

Calcium silicate hydrate is a binder that holds concrete together on its own.

The answer is (C).

5. Carbon steels are divided into four categories based on the percentages of carbon: low-carbon (less than 0.15%), mild-carbon (0.15–0.29%), medium-carbon (0.30–0.59%), and high-carbon (0.60–1.70%).

The answer is (B).

6. Yield stress is the unit tensile stress at which the stress-strain curve exhibits a well-defined increase in strain without an increase in stress. The modulus of elasticity is the slope of the initial straight-line portion of the stress-strain diagram. Toughness is the ability of a specimen to absorb energy. Ductility is the ability of steel to undergo large inelastic deformations without fracture.

The answer is (C).

7. Although the unit weight of unreinforced normal weight concrete is taken as 145 lbf/ft^3, the unit weight of reinforced normal weight concrete is usually assumed to be 150 lbf/ft^3. The weight of the beam per unit length is

$$W = bhw = \frac{(12 \text{ in})(28 \text{ in})\left(150 \dfrac{\text{lbf}}{\text{ft}^3}\right)}{\left(12 \dfrac{\text{in}}{\text{ft}}\right)^2 \left(1000 \dfrac{\text{lbf}}{\text{kip}}\right)} = 0.35 \text{ kip/ft}$$

The factored uniform load is the maximum of

$$U = 1.4D = (1.4)\left(2.0 \frac{\text{kips}}{\text{ft}} + 0.35 \frac{\text{kip}}{\text{ft}}\right)$$

$$= 3.29 \text{ kips/ft} \quad \text{[does not control]}$$

$$U = 1.2D + 1.6L$$

$$= (1.2)\left(2.0 \frac{\text{kips}}{\text{ft}} + 0.35 \frac{\text{kip}}{\text{ft}}\right) + (1.6)\left(2.4 \frac{\text{kips}}{\text{ft}}\right)$$

$$= 6.66 \text{ kips/ft} \quad (6.7 \text{ kips/ft})$$

The larger value controls, so use 6.7 kips/ft.

The answer is (A).

8. The specific weight of steel-reinforced concrete is assumed to be 150 lbf/ft^3. The weight of the beam is

$$W = bhw = \frac{(12 \text{ in})(24 \text{ in})\left(150 \dfrac{\text{lbf}}{\text{ft}^3}\right)}{\left(12 \dfrac{\text{in}}{\text{ft}}\right)^2 \left(1000 \dfrac{\text{lbf}}{\text{kip}}\right)} = 0.3 \text{ kip/ft}$$

The factored uniform load is

$$U = 1.2D + 1.6L$$

$$= (1.2)\left(1.2 \frac{\text{kips}}{\text{ft}} + 0.3 \frac{\text{kip}}{\text{ft}}\right) + (1.6)\left(0.8 \frac{\text{kip}}{\text{ft}}\right)$$

$$= 3.08 \text{ kips/ft}$$

For a uniformly loaded beam, the factored moment is maximum at the center of the beam and is

$$M_u = \frac{UL^2}{8} = \frac{\left(3.08 \dfrac{\text{kips}}{\text{ft}}\right)(30 \text{ ft})^2}{8}$$

$$= 347 \text{ ft-kips} \quad (350 \text{ ft-kips})$$

The answer is (B).

Structural Design

37 Reinforced Concrete: Beams

PRACTICE PROBLEMS

1. The span length and cross section of a reinforced concrete beam are shown. The beam is underreinforced. The concrete and reinforcing steel properties are $f'_c = 3000 \text{ lbf/in}^2$, $f_y = 40,000 \text{ lbf/in}^2$, and $A_s = 3 \text{ in}^2$.

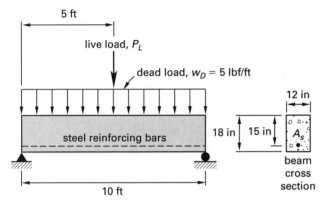

Neglecting beam self-weight and based only on the allowable moment capacity of the beam as determined using American Concrete Institute (ACI) strength design specifications, the maximum allowable live load is most nearly

(A) 23,000 lbf

(B) 29,000 lbf

(C) 35,000 lbf

(D) 50,000 lbf

2. A floor system consists of ten 30 ft long reinforced concrete beams and a continuous 5 in deck slab. (A typical section is shown for the deck and two of the beams.) Assume the beams are underreinforced.

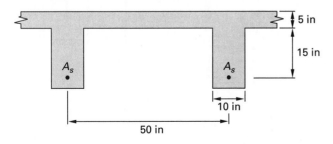

For each beam in the floor system, the ACI-specified effective top flange width is most nearly

(A) 36 in

(B) 50 in

(C) 60 in

(D) 90 in

3. The cross section of a reinforced concrete beam with tension reinforcement is shown. Assume that the beam is underreinforced.

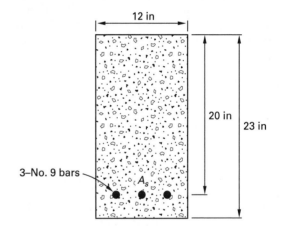

$$f'_c = 3000 \text{ lbf/in}^2$$
$$f_y = 40,000 \text{ lbf/in}^2$$
$$A_s = 3 \text{ in}^2 \quad \text{[three no. 9 bars]}$$
$$A'_s = 1 \text{ in}^2$$

If the dead load shear force in the beam is 5 kips and the live load shear force in the beam is 15 kips, then the minimum amount of shear reinforcement needed for a center-to-center stirrup spacing of 8 in based on ACI strength design is most nearly

(A) 0.10 in^2

(B) 0.12 in^2

(C) 0.14 in^2

(D) 0.18 in^2

4. The span length and cross section of a reinforced concrete beam are shown. The beam is underreinforced. The concrete and reinforcing steel properties are $f'_c = 2500$ lbf/in^2, $f_y = 50,000$ lbf/in^2, and $A_s = 3.8$ in^2.

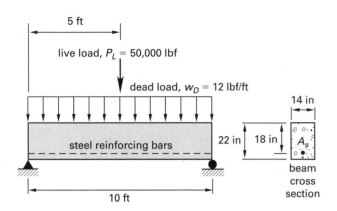

The beam supports a concentrated live load of 50,000 lbf. Neglect beam self-weight. The minimum amount of shear reinforcement required for a center-to-center stirrup spacing of 12 in under ACI strength design specifications is most nearly

(A) 0.17 in^2

(B) 0.23 in^2

(C) 0.38 in^2

(D) 0.78 in^2

5. A floor system consists of 30 reinforced concrete beams and a continuous 4 in deck slab. (A typical section is shown for the deck and two of the beams.) Assume the beams are underreinforced.

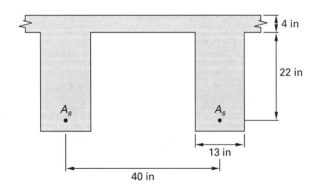

$$f'_c = 2800 \text{ lbf/in}^2$$

$$f_y = 42,000 \text{ lbf/in}^2$$

$$L = 30 \text{ ft} \quad [\text{simple span length}]$$

Assume the effective flange width for this beam is 40 in. If the area of reinforcing steel per beam is 6.00 in^2, the nominal moment capacity of each beam based on ACI strength design is most nearly

(A) 150 ft-kips

(B) 160 ft-kips

(C) 520 ft-kips

(D) 650 ft-kips

6. The span length and cross section of a reinforced concrete beam are shown. The beam is underreinforced. The concrete and reinforcing steel properties are $f'_c = 3100$ lbf/in^2, $f_y = 35,000$ lbf/in^2, and $A_s = 2.5$ in^2.

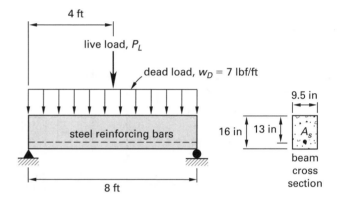

The balanced reinforcing steel ratio for this beam in accordance with ACI specifications is most nearly

(A) 0.037

(B) 0.046

(C) 0.051

(D) 0.058

7. A floor system consists of 20 reinforced concrete beams and a continuous 3 in deck slab. (A typical section is shown for the deck and two of the beams.) Assume the beams are underreinforced.

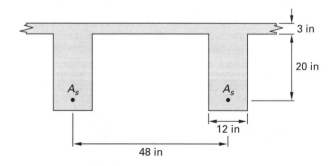

$$f'_c = 3000 \text{ lbf/in}^2$$

$$f_y = 60,000 \text{ lbf/in}^2$$

$$L = 30 \text{ ft} \quad [\text{simple span length}]$$

Assume the effective flange width for this beam is 48 in. If the area of reinforcing steel per beam is 7.25 in^2, the nominal moment capacity of each beam based on ACI strength design is most nearly

(A) 680 ft-kips

(B) 770 ft-kips

(C) 800 ft-kips

(D) 880 ft-kips

8. The cross section of a reinforced concrete beam with compression reinforcement is shown.

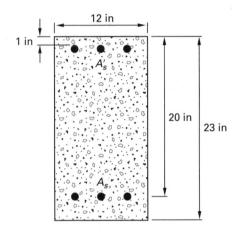

$$f'_c = 3000 \text{ lbf/in}^2$$
$$f_y = 40{,}000 \text{ lbf/in}^2$$
$$A_s = 3 \text{ in}^2$$
$$A'_s = 1 \text{ in}^2$$

The nominal moment capacity of the beam is most nearly

(A) 130 ft-kips

(B) 150 ft-kips

(C) 170 ft-kips

(D) 190 ft-kips

9. A monolithic slab-beam floor system is supported on a column grid of 18 ft on centers as shown. The dimensions of the cross section for the beams running in the north-south direction have been determined.

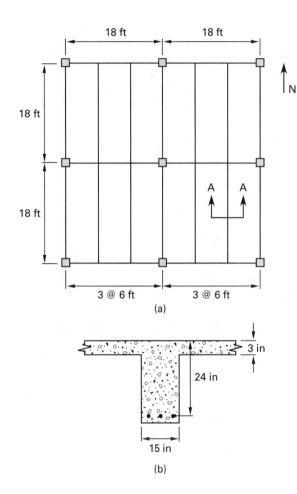

(a)

(b)

What is most nearly the effective flange width?

(A) 45 in

(B) 54 in

(C) 63 in

(D) 72 in

10. A reinforced concrete T-beam with a 30 ft span and a 30 in effective width in a floor slab system is fixed at both ends and is reinforced as shown. $f'_c = 3000$ lbf/in^2, $A_s = 3$ in^2, and $f_y = 60,000$ lbf/in^2. The stress block is within the flange.

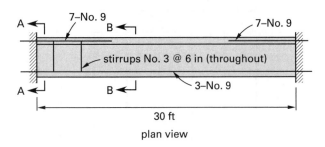

plan view

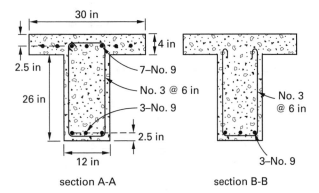

section A-A   section B-B

The nominal moment capacity in the positive moment region is most nearly

 (A) 300 ft-kips

 (B) 360 ft-kips

 (C) 390 ft-kips

 (D) 410 ft-kips

SOLUTIONS

1. The height of the stress block is

$$a = \frac{A_s f_y}{0.85 f'_c b} = \frac{(3 \text{ in}^2)\left(40{,}000 \frac{\text{lbf}}{\text{in}^2}\right)}{(0.85)\left(3000 \frac{\text{lbf}}{\text{in}^2}\right)(12 \text{ in})}$$

$$= 3.92 \text{ in}$$

For flexure, the strength reduction factor, ϕ, is 0.90.

$$M_u = \phi M_n = \phi A_s f_y \left(d - \frac{a}{2}\right)$$

$$= \frac{(0.90)(3 \text{ in}^2)\left(40{,}000 \frac{\text{lbf}}{\text{in}^2}\right)\left(15 \text{ in} - \frac{3.92 \text{ in}}{2}\right)}{12 \frac{\text{in}}{\text{ft}}}$$

$$= 117{,}350 \text{ ft-lbf}$$

The maximum bending moment occurs at midspan.

$$M_u = 1.2 M_D + 1.6 M_L$$

$$= 1.2 \frac{w_D L^2}{8} + 1.6 \frac{P_L L}{4}$$

This can be solved for the maximum allowable live load.

$$P_L = \frac{4\left(M_u - 1.2 \frac{w_D L^2}{8}\right)}{1.6 L}$$

$$= \frac{(4)\left(\begin{array}{c}(117{,}350 \text{ ft-lbf}) \\ -(1.2)\left(\frac{\left(5 \frac{\text{lbf}}{\text{ft}}\right)(10 \text{ ft})^2}{8}\right)\end{array}\right)}{(1.6)(10 \text{ ft})}$$

$$= 29{,}319 \text{ lbf} \quad (29{,}000 \text{ lbf})$$

The answer is (B).

2. The effective flange width is

$$b_e = \text{smallest} \begin{cases} \left(\frac{1}{4}\right)(\text{span length}) = \left(\frac{1}{4}\right)(30 \text{ ft})\left(12 \frac{\text{in}}{\text{ft}}\right) \\ \qquad = 90 \text{ in} \\ b_w + 16 h_f = 10 \text{ in} + (16)(5 \text{ in}) \\ \qquad = 90 \text{ in} \\ \text{beam centerline} \\ \text{spacing} = 50 \text{ in} \end{cases}$$

Therefore, the effective flange width is 50 in.

The answer is (B).

3. For shear, the capacity reduction ratio is $\phi = 0.75$. The ultimate shear force in the beam is

$$V_u = 1.2V_D + 1.6V_L$$
$$= (1.2)(5 \text{ kips}) + (1.6)(15 \text{ kips})$$
$$= 30 \text{ kips}$$

The nominal concrete shear strength is

$$V_c = 2b_w d\sqrt{f'_c}$$

$$= \dfrac{(2)(12 \text{ in})(20 \text{ in})\sqrt{3000 \dfrac{\text{lbf}}{\text{in}^2}}}{1000 \dfrac{\text{kip}}{\text{lbf}}}$$

$$= 26.29 \text{ kips}$$

$$\dfrac{\phi V_c}{2} = \dfrac{(0.75)(26.29 \text{ kips})}{2}$$

$$= 9.9 \text{ kips}$$

Since $V_u > \phi V_c/2$, shear reinforcement is required.

$$\phi V_c = (0.75)(26.29 \text{ kips}) = 19.72 \text{ kips}$$

Since $V_u = 30 \text{ kips} > \phi V_c = 19.72 \text{ kips}$, the required shear strength provided by the steel is

$$V_s = \dfrac{V_u}{\phi} - V_c = \dfrac{30 \text{ kips}}{0.75} - 26.29 \text{ kips} = 13.71 \text{ kips}$$

The required steel area is

$$A_v = \dfrac{sV_s}{f_y d} = \dfrac{(8 \text{ in})(13.71 \text{ kips})\left(1000 \dfrac{\text{lbf}}{\text{kip}}\right)}{\left(40{,}000 \dfrac{\text{lbf}}{\text{in}^2}\right)(20 \text{ in})}$$

$$= 0.1371 \text{ in}^2 \quad (0.14 \text{ in}^2)$$

Although not required for this problem, in an actual design and analysis situation, a check should be made to ensure that V_s does not exceed the ACI-allowed maximum shear reinforcement given by $V_{s,\max} = 8\sqrt{f'_c}bd$.

The answer is (C).

4. For shear, the strength reduction factor, ϕ, is 0.75. The maximum factored shear force in the beam is at either one of the supports and is

$$V_u = 1.2V_D + 1.6V_L = 1.2\dfrac{w_D L}{2} + 1.6\dfrac{P_L}{2}$$

$$= (1.2)\left(\dfrac{\left(12 \dfrac{\text{lbf}}{\text{ft}}\right)(10 \text{ ft})}{2}\right) + (1.6)\left(\dfrac{50{,}000 \text{ lbf}}{2}\right)$$

$$= 40{,}072 \text{ lbf}$$

The nominal concrete shear strength is

$$V_c = 2b_w d\sqrt{f'_c} = (2)(14 \text{ in})(18 \text{ in})\sqrt{2500 \dfrac{\text{lbf}}{\text{in}^2}}$$

$$= 25{,}200 \text{ lbf}$$

$$\dfrac{\phi V_c}{2} = \dfrac{(0.75)(25{,}200 \text{ lbf})}{2}$$

$$= 9450 \text{ lbf}$$

$$V_u > \dfrac{\phi V_c}{2}$$

Therefore, shear reinforcement is required.

In accordance with ACI specifications, the minimum required amount of shear reinforcement for a stirrup spacing of 12 in is

$$A_v = \dfrac{50b_w s}{f_y} = \dfrac{(50)(14 \text{ in})(12 \text{ in})}{50{,}000 \dfrac{\text{lbf}}{\text{in}^2}}$$

$$= 0.168 \text{ in}^2$$

The amount of shear reinforcement based on factored loading can be determined as follows.

$$V_s = A_v f_y \dfrac{d}{s}$$

$$\phi(V_c + V_s) \geq V_u$$

$$A_v = \dfrac{V_u - \phi V_c}{\phi f_y \dfrac{d}{s}}$$

$$= \dfrac{40{,}072 \text{ lbf} - (0.75)(25{,}200 \text{ lbf})}{(0.75)\left(50{,}000 \dfrac{\text{lbf}}{\text{in}^2}\right)\left(\dfrac{18 \text{ in}}{12 \text{ in}}\right)}$$

$$= 0.376 \text{ in}^2 \quad (0.38 \text{ in}^2)$$

The larger value for A_v controls. Use $A_v = 0.38 \text{ in}^2$.

The answer is (C).

5. This problem asks for the nominal moment capacity, M_n, not the allowable moment capacity, ϕM_n. Therefore, the reduction factor, ϕ, is not needed.

First, assume that each beam is rectangular with a width $b = b_e = 40 \text{ in}$. The depth of the concrete compressive stress block is

$$a = \dfrac{A_s f_y}{0.85 f'_c b}$$

$$= \dfrac{(6.00 \text{ in}^2)\left(42{,}000 \dfrac{\text{lbf}}{\text{in}^2}\right)}{(0.85)\left(2800 \dfrac{\text{lbf}}{\text{in}^2}\right)(40 \text{ in})}$$

$$= 2.65 \text{ in}$$

Since $a <$ slab depth of 4 in, the nominal moment capacity of the beam is the same as for a rectangular singly reinforced concrete beam. Using $b = b_e = 40$ in, the nominal moment capacity is

$$M_n = 0.85 f'_c ab \left(d - \frac{a}{2} \right)$$

$$= \frac{(0.85)\left(2800 \ \dfrac{\text{lbf}}{\text{in}^2} \right)(2.65 \text{ in})}{\left(12 \ \dfrac{\text{in}}{\text{ft}} \right)\left(1000 \ \dfrac{\text{lbf}}{\text{kip}} \right)}$$

$$= 520 \text{ ft-kips}$$

This can also be calculated by using the following equation in accordance with rectangular singly reinforced concrete beam theory.

$$M_n = A_s f_y \left(d - \frac{a}{2} \right)$$

Even though the problem statement assumes that the beam is underreinforced, the actual reinforcing steel ratio and its limits should always be checked in real design and analysis problems.

The answer is (C).

6. The ratio of the rectangular stress block depth to the neutral axis depth is

$$\beta_1 = 0.85 \geq \left(0.85 - 0.05 \left(\frac{f'_c - 4000}{1000} \right) \right) \geq 0.65$$

$$0.85 - (0.05)\left(\frac{f'_c - 4000}{1000} \right)$$

$$= 0.85 - (0.05)\left(\frac{3100 \ \dfrac{\text{lbf}}{\text{in}^2} - 4000 \ \dfrac{\text{lbf}}{\text{in}^2}}{1000 \ \dfrac{\text{lbf}}{\text{in}^2}} \right)$$

$$= 0.895$$

The reinforcement ratio is

$$\rho = \frac{A_s}{b d_t}$$

Expressions for A_s and d_t are needed.

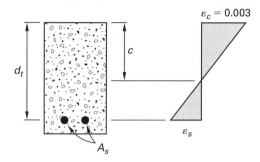

From similar triangles,

$$\frac{c}{d_t} = \frac{\varepsilon_c}{\varepsilon_c + \varepsilon_s}$$

For a balanced condition, the concrete strain is 0.003, and the steel is at yield.

$$\frac{c}{d_t} = \frac{\varepsilon_c}{\varepsilon_c + \varepsilon_y} = \frac{\varepsilon_c}{\varepsilon_c + \dfrac{f_y}{E_s}}$$

The modulus of elasticity for steel is $E_s = 29 \times 10^6$ lbf/in^2. Since $\varepsilon_c = 0.003$, for the balanced condition,

$$\frac{1}{d_t} = \frac{87{,}000}{c(87{,}000 + f_y)}$$

The required steel area can be found from the nominal moment strength.

$$M_n = 0.85 f'_c ab \left(d - \frac{a}{2} \right)$$

$$= A_s f_y \left(d - \frac{a}{2} \right)$$

$$A_s = \frac{0.85 f'_c ab}{f_y}$$

And, since $a = \beta_1 c$,

$$A_s = \frac{0.85 f'_c \beta_1 cb}{f_y}$$

Combining the expressions for ρ, $1/d_t$, and A_s, the balanced reinforcement ratio is

$$\rho_b = \left(\frac{0.85 \beta_1 f'_c}{f_y} \right)\left(\frac{87{,}000}{87{,}000 + f_y} \right)$$

Use $\beta_1 = 0.85$ since this is its maximum allowed value.

$$\rho_b = \left(\frac{0.85\beta_1 f'_c}{f_y}\right)\left(\frac{87{,}000\ \frac{\text{lbf}}{\text{in}^2}}{87{,}000\ \frac{\text{lbf}}{\text{in}^2} + f_y}\right)$$

$$= \left(\frac{(0.85)(0.85)\left(3100\ \frac{\text{lbf}}{\text{in}^2}\right)}{35{,}000\ \frac{\text{lbf}}{\text{in}^2}}\right)$$

$$\times \left(\frac{87{,}000\ \frac{\text{lbf}}{\text{in}^2}}{87{,}000\ \frac{\text{lbf}}{\text{in}^2} + 35{,}000\ \frac{\text{lbf}}{\text{in}^2}}\right)$$

$$= 0.0456 \quad (0.046)$$

The answer is (B).

7. This problem asks for the nominal moment capacity, M_n, not the allowable moment capacity, ϕM_n. Therefore, the reduction factor, ϕ, is not needed.

The depth of the concrete compressive stress block must be checked to see whether or not it exceeds the 3 in deck thickness. If the depth of this compressive stress block exceeds the deck thickness, then each beam is a T-beam and T-beam formulas apply for determination of the nominal moment capacity. If, however, the depth of the concrete compressive stress block does not exceed the deck thickness, then each beam is a rectangular beam and rectangular beam formulas apply for determination of the nominal moment capacity.

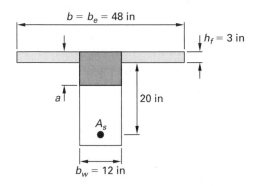

First, assume that each beam is rectangular with a width of $b = b_e = 48$ in. The depth of the concrete compressive stress block is

$$a = \frac{A_s f_y}{0.85 f'_c b} = \frac{(7.25\ \text{in}^2)\left(60{,}000\ \frac{\text{lbf}}{\text{in}^2}\right)}{(0.85)\left(3000\ \frac{\text{lbf}}{\text{in}^2}\right)(48\ \text{in})}$$

$$= 3.55\ \text{in}$$

Since $a >$ slab depth of 3 in, the beam is a T-beam.

Since $a = 3.55$ in was found by assuming that the beam was a rectangular beam with width $b = 48$ in, this depth only indicates whether or not the beam is a T-beam and is not the correct depth for determining the nominal moment capacity. The correct depth is now found by applying the concepts of static equilibrium to the beam.

To find the correct depth, a, sum horizontal forces in the T-beam to show that the upper (above the neutral axis) compressive concrete stress block force is equal to the lower (below the neutral axis) maximum tensile force sustained by the reinforcing bars. By dividing the entire concrete compressive stress block section into three parts (a rectangular part and two overhanging flanges), depth a can be found.

A_f is the area of overhanging flanges. A_c is the total area of concrete compressive stress block. A_r is the area of concrete compressive stress block in the rectangular part of the T-beam between the overhanging flanges.

$$A_f = (b_e - b_w)h_f$$

$$= (48\ \text{in} - 12\ \text{in})(3\ \text{in})$$

$$= 108\ \text{in}^2$$

From equilibrium of horizontal forces,

$$0.85 f'_c A_c = A_s f_y$$

$$A_c = \frac{A_s f_y}{0.85 f'_c}$$

$$= \frac{(7.25\ \text{in}^2)\left(60{,}000\ \frac{\text{lbf}}{\text{in}^2}\right)}{(0.85)\left(3000\ \frac{\text{lbf}}{\text{in}^2}\right)}$$

$$= 170.59\ \text{in}^2$$

$$A_r = b_w a = A_c - A_f$$

$$a = \frac{A_c - A_f}{b_w}$$

$$= \frac{170.59\ \text{in}^2 - 108\ \text{in}^2}{12\ \text{in}}$$

$$= 5.22\ \text{in}$$

Alternatively, a redefined stress block depth could be used.

$$a = \frac{A_s f_y}{0.85 f'_c b_w} - \frac{h_f(b_e - b_w)}{b_w}$$

$$= \frac{(7.25\ \text{in}^2)\left(60{,}000\ \frac{\text{lbf}}{\text{in}^2}\right)}{(0.85)\left(3000\ \frac{\text{lbf}}{\text{in}^2}\right)(12\ \text{in})} - \frac{(3\ \text{in})(48\ \text{in} - 12\ \text{in})}{12\ \text{in}}$$

$$= 5.22\ \text{in}$$

The nominal moment capacity of the T-beam is

$$M_n = 0.85f'_c h_f (b_e - b_w) \left(d - \frac{h_f}{2} \right)$$
$$+ 0.85f'_c a b_w \left(d - \frac{a}{2} \right)$$

$$= \frac{\begin{matrix} (0.85)\left(3000 \; \frac{\text{lbf}}{\text{in}^2}\right)(3 \text{ in}) \\ \times (48 \text{ in} - 12 \text{ in})\left(23 \text{ in} - \frac{3 \text{ in}}{2}\right) \\ + (0.85)\left(3000 \; \frac{\text{lbf}}{\text{in}^2}\right)(5.22 \text{ in}) \\ \times (12 \text{ in})\left(23 \text{ in} - \frac{5.22 \text{ in}}{2}\right) \end{matrix}}{\left(12 \; \frac{\text{in}}{\text{ft}}\right)\left(1000 \; \frac{\text{lbf}}{\text{kip}}\right)}$$

$$= 765 \text{ ft-kips} \quad (770 \text{ ft-kips})$$

Even though the problem statement assumes that the beam is underreinforced, the actual reinforcing steel ratio and its limits should always be checked in real design and analysis problems.

The answer is (B).

8. Determine whether compression steel yields.

$$A_s - A'_s = 3 \text{ in}^2 - 1 \text{ in}^2$$
$$= 2 \text{ in}^2$$

$$\frac{0.85\beta_1 f'_c d' b}{f_y} \left(\frac{87,000}{87,000 - f_y} \right)$$

$$= \frac{(0.85)(0.85)\left(3000 \; \frac{\text{lbf}}{\text{in}^2}\right)(1 \text{ in})(12 \text{ in})}{40,000 \; \frac{\text{lbf}}{\text{in}^2}}$$

$$\times \left(\frac{87,000 \; \frac{\text{lbf}}{\text{in}^2}}{87,000 \; \frac{\text{lbf}}{\text{in}^2} - 40,000 \; \frac{\text{lbf}}{\text{in}^2}} \right)$$

$$= 1.2 \text{ in}^2$$

Compare both sides of the equation.

$$A_s - A'_s \geq \frac{0.85\beta_1 f'_c d' b}{f_y} \left(\frac{87,000}{87,000 - f_y} \right)$$

$$2 \text{ in}^2 \geq 1.2 \text{ in}^2$$

Therefore, compression steel yields.

Calculate the depth of the concrete compressive stress block.

$$a = \frac{(A_s - A'_s)f_y}{0.85f'_c b} = \frac{(3 \text{ in}^2 - 1 \text{ in}^2)\left(40,000 \; \frac{\text{lbf}}{\text{in}^2}\right)}{(0.85)\left(3000 \; \frac{\text{lbf}}{\text{in}^2}\right)(12 \text{ in})}$$

$$= 2.61 \text{ in}$$

The nominal moment strength is

$$M_n = f_y \left((A_s - A'_s)\left(d - \frac{a}{2} \right) + A'_s(d - d') \right)$$

$$= \frac{\left(40,000 \; \frac{\text{lbf}}{\text{in}^2}\right)\left(\begin{matrix} (3 \text{ in}^2 - 1 \text{ in}^2)\left(20 \text{ in} - \frac{2.61 \text{ in}}{2}\right) \\ + (1 \text{ in}^2)(20 \text{ in} - 1 \text{ in}) \end{matrix} \right)}{\left(12 \; \frac{\text{in}}{\text{ft}}\right)\left(1000 \; \frac{\text{lbf}}{\text{kip}}\right)}$$

$$= 187.95 \text{ ft-kips} \quad (190 \text{ ft-kips})$$

The answer is (D).

9. The effective width of the flange is determined as the smallest of

$$b_e = \text{smallest} \begin{cases} \left(\frac{1}{4}\right)(\text{span length}) = \left(\frac{1}{4}\right)(18 \text{ ft})\left(12 \; \frac{\text{in}}{\text{ft}}\right) \\ \qquad\qquad = 54 \text{ in} \\ b_w + 16h_f = 15 \text{ in} + (16)(3 \text{ in}) \\ \qquad\qquad = 63 \text{ in} \\ \begin{matrix} \text{beam centerline} \\ \text{spacing} \end{matrix} = (6 \text{ ft})\left(12 \; \frac{\text{in}}{\text{ft}}\right) \\ \qquad\qquad = 72 \text{ in} \end{cases}$$

Therefore, the effective flange width is 54 in.

The answer is (B).

10. The positive moment region for slabs is the central region, so cross-section B-B is applicable. Since the stress block is within the flange, $a < h_f$, and the T-beam can be analyzed like a rectangular beam. The distance from the extreme compression fibers to the extreme tensile steel is

$$d = 4 \text{ in} + 26 \text{ in} - 2.5 \text{ in}$$
$$= 27.5 \text{ in}$$

Determine the height of the stress block.

$$a = \frac{A_s f_y}{0.85f'_c b_e} = \frac{(3 \text{ in}^2)\left(60 \; \frac{\text{kips}}{\text{in}^2}\right)}{(0.85)\left(3 \; \frac{\text{kips}}{\text{in}^2}\right)(30 \text{ in})} = 2.35 \text{ in}$$

The moment capacity can be calculated from either the steel properties (with $b = b_e$) or from the concrete properties. Using the steel properties,

$$M_n = A_s f_y \left(d - \frac{a}{2} \right)$$

$$= \frac{(3 \text{ in}^2) \left(60 \ \dfrac{\text{kips}}{\text{in}^2} \right) \left(27.5 \text{ in} - \dfrac{2.35 \text{ in}}{2} \right)}{12 \ \dfrac{\text{in}}{\text{ft}}}$$

$$= 394.9 \text{ ft-kips} \quad (390 \text{ ft-kips})$$

Using the concrete properties,

$$M_n = 0.85 f'_c a b_e \left(d - \frac{a}{2} \right)$$

$$= \frac{\begin{aligned} (0.85) &\left(3 \ \dfrac{\text{kips}}{\text{in}^2} \right) (2.35 \text{ in})(30 \text{ in}) \\ &\times \left(27.5 \text{ in} - \dfrac{2.35 \text{ in}}{2} \right) \end{aligned}}{12 \ \dfrac{\text{in}}{\text{ft}}}$$

$$= 394.9 \text{ ft-kips} \quad (390 \text{ ft-kips})$$

The answer is (C).

Structural Design

38 Reinforced Concrete: Columns

PRACTICE PROBLEMS

1. For the short, concentrically loaded round spiral column shown, the applied axial dead load is 150 kips, and the applied axial live load is 350 kips.

$$f'_c = 4000 \text{ lbf/in}^2$$
$$f_y = 60,000 \text{ lbf/in}^2$$

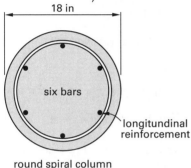

round spiral column
(cross section)

Assuming that the longitudinal reinforcing bars are all the same size, the minimum required size of each longitudinal reinforcing bar is

(A) no. 7

(B) no. 8

(C) no. 9

(D) no. 10

2. For the short, concentrically loaded square tied column shown, the applied axial dead load is 150 kips, and the applied axial live load is 250 kips.

$$f'_c = 4000 \text{ lbf/in}^2$$
$$f_y = 60,000 \text{ lbf/in}^2$$

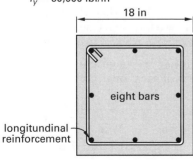

square tied column
(cross section)

Assuming that the longitudinal reinforcing bars are all the same size, the minimum required size of each longitudinal reinforcing bar is

(A) no. 3

(B) no. 4

(C) no. 5

(D) no. 6

3. A reinforced concrete tied column is subjected to a design axial compression force of 1090 kips that is concentrically applied. Slenderness effects are negligible, and the column is to be designed using ACI 318. Given a specified compressive strength of 5000 psi, grade 60 rebars, and a specified longitudinal steel ratio of 0.02, what is most nearly the width of the sides of the smallest square column that will support the load?

(A) 12 in

(B) 16 in

(C) 20 in

(D) 24 in

4. A 16 in (gross dimension) square, tied column must carry 220 kip dead and 250 kip live loads. The dead load includes the column self-weight. The column is not exposed to any moments. Sidesway is prevented at the top, and slenderness effects are to be disregarded. The concrete compressive strength is 4000 lbf/in², and the steel tensile yield strength is 60,000 lbf/in². The longitudinal reinforcement of this column is most nearly

(A) 0.028

(B) 0.061

(C) 0.092

(D) 0.11

5. An 18 in square tied column is reinforced with 12 no. 9 grade 60 bars and has a concrete compressive strength of 4000 lbf/in². The column, which is braced against sidesway, has an unsupported height of 9 ft and supports axial load only without end moments. What is most nearly the design axial load capacity?

(A) 930 kips

(B) 970 kips

(C) 1800 kips

(D) 1900 kips

6. The short spiral column shown uses eight no. 8 bars. Assume the loading has a low eccentricity. $f'_c = 3500$ lbf/in^2, and $f_y = 40{,}000$ lbf/in^2.

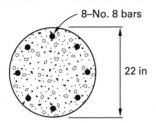

The design strength is most nearly

(A) 810 kips

(B) 870 kips

(C) 890 kips

(D) 8100 kips

SOLUTIONS

1. Determine the amount of reinforcing steel required by the minimum required reinforcement ratio, ρ_g, of 0.01. The minimum area of reinforcing steel required is

$$A_s = \rho_g A_g = (0.01)\left(\frac{\pi(18 \text{ in})^2}{4}\right)$$

$$= 2.54 \text{ in}^2$$

Determine the required amount of reinforcing steel based on the factored axial load, P_u.

$$P_u = 1.2P_D + 1.6P_L$$
$$= (1.2)(150 \text{ kips}) + (1.6)(350 \text{ kips})$$
$$= 740 \text{ kips}$$

The nominal axial compressive load capacity is

$$P_n = 0.85P_o$$
$$= (0.85)(0.85f'_c A_{\text{concrete}} + f_y A_s)$$
$$= (0.85)(0.85f'_c(A_g - A_s) + f_y A_s)$$

It is required that $\phi P_n \geq P_u$. For axial compression with spiral reinforcement, $\phi = 0.70$. Setting $\phi P_n = P_u$ and solving for the area of longitudinal reinforcing steel gives

$$A_s = \frac{\dfrac{P_u}{0.85\phi} - 0.85f'_c A_g}{f_y - 0.85f'_c}$$

$$= \frac{\begin{aligned}&\dfrac{(740 \text{ kips})\left(1000 \, \dfrac{\text{lbf}}{\text{kip}}\right)}{(0.85)(0.70)} \\ &\quad - (0.85)\left(4000 \, \dfrac{\text{lbf}}{\text{in}^2}\right)\left(\dfrac{\pi(18 \text{ in})^2}{4}\right)\end{aligned}}{60{,}000 \, \dfrac{\text{lbf}}{\text{in}^2} - (0.85)\left(4000 \, \dfrac{\text{lbf}}{\text{in}^2}\right)}$$

$$= 6.69 \text{ in}^2$$

$A_s = 6.69$ in^2 is greater than $A_s = 2.54$ in^2 based on the minimum allowed reinforcement ratio, $\rho_g = 0.01$. The minimum required area of reinforcement is $A_s = 6.69$ in^2.

The column has six longitudinal reinforcing bars. The required area of each longitudinal reinforcing bar is

$$A = \frac{A_s}{n_{\text{bars}}} = \frac{6.69 \text{ in}^2}{6 \text{ bars}}$$

$$= 1.11 \text{ in}^2/\text{bar}$$

A bar area of 1.11 in^2 is satisfied by a no. 10 bar, which has a nominal area of 1.27 in^2.

In an actual design and analysis situation, a check should also be made to see that the actual longitudinal reinforcement ratio does not exceed the maximum allowable ratio of 0.08.

The answer is (D).

2. The minimum required reinforcement ratio, ρ_g, is 0.01. The minimum area of reinforcing steel required is

$$A_s = \rho_g A_g = (0.01)(18 \text{ in})^2$$
$$= 3.24 \text{ in}^2$$

Determine the required amount of reinforcing steel based on the factored axial load, P_u.

$$P_u = 1.2P_D + 1.6P_L$$
$$= (1.2)(150 \text{ kips}) + (1.6)(250 \text{ kips})$$
$$= 580 \text{ kips}$$

The nominal axial compressive load capacity is

$$P_n = 0.80P_o$$
$$= (0.80)(0.85f'_c A_{\text{concrete}} + f_y A_s)$$
$$= (0.80)(0.85f'_c (A_g - A_s) + f_y A_s)$$

It is required that $\phi P_n \geq P_u$. For axial compression with tied reinforcement, $\phi = 0.65$. Setting $\phi P_n = P_u$ and solving for the area of longitudinal reinforcing steel gives

$$A_s = \frac{\dfrac{P_u}{0.80\phi} - 0.85f'_c A_g}{f_y - 0.85f'_c}$$

$$= \frac{\dfrac{(580 \text{ kips})\left(1000 \dfrac{\text{lbf}}{\text{kip}}\right)}{(0.80)(0.65)} - (0.85)\left(4000 \dfrac{\text{lbf}}{\text{in}^2}\right)(18 \text{ in})^2}{60{,}000 \dfrac{\text{lbf}}{\text{in}^2} - (0.85)\left(4000 \dfrac{\text{lbf}}{\text{in}^2}\right)}$$

$$= 0.244 \text{ in}^2$$

$A_s = 0.244 \text{ in}^2$ is less than $A_s = 3.24 \text{ in}^2$ based on the minimum allowed reinforcement ratio, $\rho_g = 0.01$. Therefore, the minimum required area of reinforcement is $A_s = 3.24 \text{ in}^2$.

The square tied column has eight uniformly sized longitudinal reinforcing bars. The required area of each longitudinal reinforcing bar is

$$A = \frac{A_s}{n_{\text{bars}}} = \frac{3.24 \text{ in}^2}{8 \text{ bars}}$$
$$= 0.405 \text{ in}^2/\text{bar}$$

A bar area of 0.405 in^2 corresponds to a no. 6 bar, which has a nominal area of 0.44 in^2.

In an actual design and analysis situation, a check should also be made to see that the actual longitudinal reinforcement ratio does not exceed the maximum allowable ratio of 0.08.

The answer is (D).

3. For a tied column, $\phi = 0.65$. For a concentrically loaded tied column, the design strength is given by

$$\phi P_{n,\text{max}} = 0.80\phi(0.85f'_c(A_g - A_{st}) + f_y A_{st})$$

For a specified longitudinal steel ratio of 0.02,

$$A_{st} = \rho_g A_g = 0.02A_g$$

Substituting gives

$$1090 \text{ kips} = \phi P_{n,\text{max}}$$
$$= (0.80)(0.65)$$
$$\times \left(\begin{array}{c} (0.85)\left(5 \dfrac{\text{kips}}{\text{in}^2}\right)(A_g - 0.02A_g) \\ + \left(60 \dfrac{\text{kips}}{\text{in}^2}\right)(0.02A_g) \end{array} \right)$$
$$A_g = 391 \text{ in}^2$$
$$b = \sqrt{A_g} = \sqrt{391 \text{ in}^2}$$
$$= 19.8 \text{ in} \quad (20 \text{ in})$$

The answer is (C).

4. Determine the design load, P_u.

$$P_u = 1.2P_D + 1.6P_L$$
$$= (1.2)(220 \text{ kips}) + (1.6)(250 \text{ kips})$$
$$= 664 \text{ kips}$$

The gross column area is

$$A_g = (16 \text{ in})(16 \text{ in}) = 256 \text{ in}^2$$

For a tied column, $\phi = 0.65$. The capacity is

$$\phi P_{n,\text{max}} = 0.80\phi(0.85f'_c(A_g - A_{st}) + f_y A_{st})$$
$$= (0.80)(0.65)$$
$$\times \left(\begin{array}{c} (0.85)\left(4 \dfrac{\text{kips}}{\text{in}^2}\right)(256 \text{ in}^2 - A_{st}) \\ + \left(60 \dfrac{\text{kips}}{\text{in}^2}\right)A_{st} \end{array} \right)$$
$$= 452.6 \text{ kips} + \left(29.4 \dfrac{\text{kips}}{\text{in}^2}\right)A_{st}$$

Structural Design

The design criterion is

$$\phi P_{n,\text{max}} \geq P_u$$

$$664 \text{ kips} = 452.6 \text{ kips} + \left(29.4 \ \frac{\text{kips}}{\text{in}^2}\right) A_{st}$$

$$A_{st} = 7.19 \text{ in}^2$$

$$\rho = \frac{A_{st}}{A_g} = \frac{7.19 \text{ in}^2}{256 \text{ in}^2} = 0.028$$

Check the limits.

$$\rho_{\text{min}} < \rho < \rho_{\text{max}}$$
$$0.01 < 0.028 < 0.08 \quad [\text{OK}]$$

The answer is (A).

5. The effective length factor for a column braced against sidesway is

$$K = 1.0$$

The radius of gyration is

$$r = 0.288h = (0.288)(18 \text{ in})$$
$$= 5.2 \text{ in}$$

The slenderness ratio is

$$\frac{KL}{r} = \frac{(1.0)(9 \text{ ft})\left(12 \ \frac{\text{in}}{\text{ft}}\right)}{5.2 \text{ in}}$$
$$= 20.8$$

The gross area of the column is

$$A_g = b^2 = (18 \text{ in})^2 = 324 \text{ in}^2$$

The area of longitudinal steel reinforcement is

$$A_{st} = NA_b = (12)(1 \text{ in}^2) = 12 \text{ in}^2$$

Since $M_1 = 0$, the column is a short column. $\phi = 0.65$ for a tied column. The design axial load capacity is

$$\phi P_n = 0.80\phi\big(0.85f'_c(A_g - A_{st}) + A_{st}f_y\big)$$

$$= (0.80)(0.65) \begin{pmatrix} (0.85)\left(4 \ \dfrac{\text{kips}}{\text{in}^2}\right) \\ \times (324 \text{ in}^2 - 12 \text{ in}^2) \\ + (12 \text{ in}^2)\left(60 \ \dfrac{\text{kips}}{\text{in}^2}\right) \end{pmatrix}$$

$$= 926 \text{ kips} \quad (930 \text{ kips})$$

The answer is (A).

6. The gross area of the column is

$$A_g = \frac{\pi D_g^2}{4} = \frac{\pi (22 \text{ in})^2}{4} = 380.1 \text{ in}^2$$

Since there are eight no. 8 bars, the area of steel is

$$A_{st} = (8)(0.79 \text{ in}^2) = 6.32 \text{ in}^2$$

$\phi = 0.70$ for spiral columns. The design strength is then

$$\phi P_n = 0.85\phi\big(0.85f'_c(A_g - A_{st}) + A_{st}f_y\big)$$
$$= (0.85)(0.70)$$

$$\times \left(\frac{(0.85)\left(3500 \ \dfrac{\text{lbf}}{\text{in}^2}\right)(380.1 \text{ in}^2 - 6.32 \text{ in}^2) + (6.32 \text{ in}^2)\left(40{,}000 \ \dfrac{\text{lbf}}{\text{in}^2}\right)}{1000 \ \dfrac{\text{lbf}}{\text{kip}}} \right)$$

$$= 812.1 \text{ kips} \quad (810 \text{ kips})$$

The answer is (A).

39 Reinforced Concrete: Slabs

PRACTICE PROBLEMS

1. A two-way slab supported on a column grid without the use of beams is known as a

(A) flat slab

(B) flat plate

(C) drop panel

(D) waffle slab

2. All of the following are conditions that must be satisfied in order to use the simplified method for computing shear and moments in one-way slabs, given in ACI 318 Sec. 8.3.3, EXCEPT

(A) There are two or more spans.

(B) Spans are approximately equal, with the longer of two adjacent spans not longer than the shorter by more than 20%.

(C) The loads are uniformly distributed.

(D) The ratio of live to dead loads is no more than 5.

SOLUTIONS

1. A two-way slab supported on a column grid without the use of beams is known as a flat plate.

The answer is (B).

2. Shear and moments in one-way slabs can be computed using the simplified method specified in ACI 318 Sec. 8.3.3 when the following conditions are satisfied: (a) There are two or more spans. (b) Spans are approximately equal, with the longer of two adjacent spans not longer than the shorter by more than 20%. (c) The loads are uniformly distributed. (d) The ratio of live to dead loads is no more than 3. (e) The slab has a uniform thickness.

The answer is (D).

40 Reinforced Concrete: Walls

PRACTICE PROBLEMS

1. According to ACI 318 Sec. 14.3.3, the minimum horizontal reinforcement for nonbearing walls is

(A) 0.0012 times the gross concrete area for deformed no. 5 bars or smaller and $f_y \geq 60{,}000$ psi

(B) 0.0015 times the gross concrete area for other deformed bars

(C) 0.0020 times the gross concrete area for smooth or deformed welded wire reinforcement not larger than W31 or D31

(D) 0.01 times the gross concrete area

SOLUTIONS

1. ACI 318 Sec. 14.3.3 gives the minimum horizontal reinforcement for nonbearing walls as (a) 0.0020 times the gross concrete area for deformed no. 5 bars or smaller and $f_y \geq 60{,}000$ psi; (b) 0.0025 times the gross concrete area for other deformed bars; and (c) 0.0020 times the gross concrete area for smooth or deformed welded wire reinforcement not larger than W31 or D31.

The answer is (C).

Structural Design

41 Reinforced Concrete: Footings

PRACTICE PROBLEMS

1. The critical sections are a distance, d, from the face of the column in

    (A) double-action shear

    (B) one-way shear

    (C) punching shear

    (D) two-way shear

SOLUTIONS

1. For one-way shear (also known as single-action shear and wide-beam shear), the critical sections are a distance, d, from the face of the column.

The answer is (B).

42 Structural Steel: Beams

PRACTICE PROBLEMS

1. A 25 ft long steel beam is loaded uniformly (live) at 4 kips/ft. Loading due to self-weight is negligible, and there is adequate lateral support provided to the beam. The required plastic section modulus for a W12 shape using grade-50 steel is most nearly

(A) 60 in^3

(B) 95 in^3

(C) 110 in^3

(D) 120 in^3

2. A W-shaped beam has a warping constant of 3450 in^6. The moment of inertia about the strong axis is 45.1 in^4, and its elastic section modulus about the weak axis is 99.1 in^3. The effective radius of gyration of the compression flange is most nearly

(A) 2.0 in

(B) 3.0 in

(C) 4.0 in

(D) 5.0 in

3. An I-shaped steel beam is built up with 0.300 in thick plates. The material's yield strength is 50 ksi. If the height of the web is 18 in, the available shear strength is most nearly

(A) 140 kips

(B) 150 kips

(C) 180 kips

(D) 190 kips

SOLUTIONS

1. Determine the required moment strength.

$$M_u = 1.6 M_L = 1.6 \frac{w_u L^2}{8}$$

$$= (1.6) \left(\frac{\left(4 \, \frac{\text{kips}}{\text{ft}} \right) (25 \text{ ft})^2}{8} \right)$$

$$= 499 \text{ ft-kips}$$

The required plastic moment is equal to the ultimate moment.

The required plastic section modulus is

$$\phi M_p = \phi F_y Z_x$$

$$Z_x = \frac{\phi M_p}{\phi F_y}$$

$$= \frac{(0.90)(499 \text{ ft-kips}) \left(12 \, \frac{\text{in}}{\text{ft}} \right)}{(0.90) \left(50 \, \frac{\text{kips}}{\text{in}^2} \right)}$$

$$= 119.76 \text{ in}^3 \quad (120 \text{ in}^3)$$

The answer is (D).

2. The effective radius of gyration is

$$r_{ts} = \sqrt[4]{\frac{I_y C_w}{S_x^2}}$$

$$= \sqrt[4]{\frac{(45.1 \text{ in}^4)(3450 \text{ in}^6)}{(99.1 \text{ in}^3)^2}}$$

$$= 1.995 \text{ in} \quad (2.0 \text{ in})$$

The answer is (A).

3. Find the area of the web.

$$A_w = dt_w = (h + 2t_f)t_w$$
$$= \left(18 \text{ in} + (2)(0.300 \text{ in})\right)(0.300 \text{ in})$$
$$= 5.58 \text{ in}^2$$

Determine which shear equation is applicable.

$$\frac{h}{t_w} = \frac{18 \text{ in}}{0.300 \text{ in}} = 60.0$$

$$\frac{418}{\sqrt{F_y}} = \frac{418}{\sqrt{50 \dfrac{\text{kips}}{\text{in}^2}}} = 59.1$$

$$\frac{522}{\sqrt{F_y}} = \frac{522}{\sqrt{50 \dfrac{\text{kips}}{\text{in}^2}}} = 73.8$$

$$\frac{418}{\sqrt{F_y}} < \frac{h}{t_w} \leq \frac{522}{\sqrt{F_y}}$$

Therefore, the available shear strength is

$$\phi V_n = \phi(0.6F_y)dt_w\left[\frac{418}{(h/t_w)\sqrt{F_y}}\right]$$
$$= (0.90)\left((0.6)\left(50 \frac{\text{kips}}{\text{in}^2}\right)\right)(5.58 \text{ in}^2)$$
$$\times \left(\frac{418}{\dfrac{18.0 \text{ in}}{0.300 \text{ in}}\sqrt{50 \dfrac{\text{kips}}{\text{in}^2}}}\right)$$
$$= 148.4 \text{ kips} \quad (150 \text{ kips})$$

The answer is (B).

43 Structural Steel: Columns

PRACTICE PROBLEMS

1. A steel compression member has a fixed support at one end and a frictionless ball joint support at the other as shown. The total applied design load consists of a dead load of 7 kips (which includes the weight of the member) and an unspecified live load. Design (not theoretical) effective lengths are to be used.

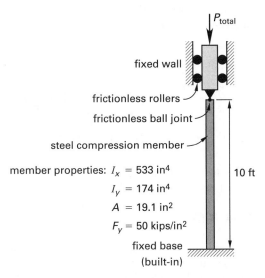

This compression member is controlled by which type of buckling?

- (A) local
- (B) torsional
- (C) inelastic
- (D) elastic

2. A long column member has one end built-in and the other end pinned. The column is loaded in compression evenly until buckling occurs. Which statement about the column after buckling is true?

- (A) The column experiences maximum deflection on its midpoint.
- (B) The maximum deflection point is closer to the pinned end than the built-in end.
- (C) The deflection curve is S-shaped.
- (D) The column experiences no deflection under buckling.

3. A solid steel column with a fixed bottom support and free upper end is concentrically loaded. Material and geometric properties are shown.

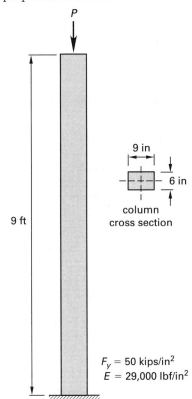

The available axial compressive design stress is most nearly

- (A) 13 kips/in^2
- (B) 18 kips/in^2
- (C) 29 kips/in^2
- (D) 39 kips/in^2

4. A steel column is built-in at one end and free to translate and rotate at the other end. The column uses a 12 ft long W12 × 45 beam. If the yield strength of the steel is 50 kips/in^2, the available design stress in the column is most nearly

- (A) 7.3 kips/in^2
- (B) 9.0 kips/in^2
- (C) 9.7 kips/in^2
- (D) 10 kips/in^2

SOLUTIONS

1. The effective column length factor for design use about both the x-axis and y-axis is 0.80.

The unbraced length of the compression member is the same about the x-axis and the y-axis.

$$
\begin{aligned}
L_x = L_y &= (10 \text{ ft})\left(12 \; \frac{\text{in}}{\text{ft}}\right) \\
&= 120 \text{ in}
\end{aligned}
$$

The radius of gyration about the x-axis is

$$
\begin{aligned}
r_x &= \sqrt{\frac{I_x}{A}} = \sqrt{\frac{533 \text{ in}^4}{19.1 \text{ in}^2}} \\
&= 5.28 \text{ in}
\end{aligned}
$$

The slenderness ratio about the x-axis is

$$
\begin{aligned}
\text{slenderness ratio}_x &= \frac{K_x L_x}{r_x} = \frac{(0.80)(120 \text{ in})}{5.28 \text{ in}} \\
&= 18.2
\end{aligned}
$$

The radius of gyration about the y-axis is

$$
\begin{aligned}
r_y &= \sqrt{\frac{I_y}{A}} = \sqrt{\frac{174 \text{ in}^4}{19.1 \text{ in}^2}} \\
&= 3.02 \text{ in}
\end{aligned}
$$

The slenderness ratio about the y-axis is

$$
\begin{aligned}
\text{slenderness ratio}_y &= \frac{K_y L_y}{r_y} = \frac{(0.80)(120 \text{ in})}{3.02 \text{ in}} \\
&= 31.8
\end{aligned}
$$

The larger slenderness ratio controls, so use 31.8.

The modulus of elasticity, E, is 29,000 kips/in^2. The limiting slenderness ratio is

$$
4.71\sqrt{\frac{E}{F_y}} = 4.71\sqrt{\frac{29,000 \; \frac{\text{kips}}{\text{in}^2}}{50 \; \frac{\text{kips}}{\text{in}^2}}} = 113.4
$$

Since 31.8 is less than 113.4, the column fails by inelastic buckling.

The answer is (C).

2. The built-in end of the column is completely fixed against rotation and translation. The pinned end of the column is free to rotate but is fixed against translation. Therefore, the pinned end of the column experiences more rotational deflection, and the maximum deflection point is closer to the pinned end of the column.

The answer is (B).

3. Since the unsupported length of the column is the same about both the strong and weak axes, the largest slenderness ratio results from bending about the weak axis. Therefore, the least radius of gyration applies.

$$
I_{\text{weak}} = r^2 A
$$

$$
\begin{aligned}
r &= \sqrt{\frac{I_{\text{weak}}}{A}} = \sqrt{\frac{\dfrac{bh^3}{12}}{bh}} \\
&= \sqrt{\frac{\dfrac{(9 \text{ in})(6 \text{ in})^3}{12}}{(9 \text{ in})(6 \text{ in})}} \\
&= 1.73 \text{ in}
\end{aligned}
$$

The effective length factor, K, is 2.10 for the given column end support conditions (fixed-free). Therefore, the slenderness ratio is

$$
\begin{aligned}
\text{slenderness ratio} &= \frac{KL}{r} \\
&= \frac{(2.10)(9 \text{ ft})\left(12 \; \frac{\text{in}}{\text{ft}}\right)}{1.73 \text{ in}} \\
&= 130.94 \quad (131)
\end{aligned}
$$

The available column strength for a slenderness ratio of 131 is 13.2 kips/in^2 (13 kips/in^2).

The answer is (A).

4. The y-axis has the smallest radius of gyration. The radius of gyration about the y-axis is 1.95 in for a W12 × 45 beam. The effective length factor is 2.10 when one end is fixed and one end is free. Use these values to determine the slenderness ratio.

$$
\begin{aligned}
\text{slenderness ratio} &= \frac{KL}{r} \\
&= \frac{(2.10)(12 \text{ ft})\left(12 \; \frac{\text{in}}{\text{ft}}\right)}{1.95 \text{ in}} \\
&= 155.1
\end{aligned}
$$

Determine the available design stress. The modulus of elasticity, E, is 29,000 kips/in^2.

$$
\begin{aligned}
4.71\sqrt{\frac{E}{F_y}} &= 4.71\sqrt{\frac{29,000 \; \frac{\text{kips}}{\text{in}^2}}{50 \; \frac{\text{kips}}{\text{in}^2}}} \\
&= 113.4
\end{aligned}
$$

The slenderness ratio is greater than 113.4. Therefore, the elastic buckling stress is

$$F_e = \frac{\pi^2 E}{\left(\dfrac{KL}{r}\right)^2}$$

$$= \frac{\pi^2 \left(29{,}000 \ \dfrac{\text{kips}}{\text{in}^2}\right)}{155.1^2}$$

$$= 11.90 \ \text{kips/in}^2$$

The available design stress is

$$F_{\text{cr}} = 0.877 F_e = (0.877)\left(11.90 \ \frac{\text{kips}}{\text{in}^2}\right)$$

$$= 10.44 \ \text{kips/in}^2 \quad (10 \ \text{kips/in}^2)$$

The answer is (D).

Structural
Design

44 Structural Steel: Tension Members

PRACTICE PROBLEMS

1. A bolted steel tension member is shown.

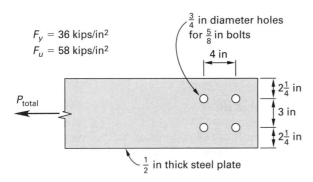

What is most nearly the effective net area in tension for this plate?

- (A) 2.3 in²
- (B) 2.9 in²
- (C) 3.4 in²
- (D) 3.8 in²

2. A steel tension member is 5 in long and ½ in thick. There are two holes in the bar. The holes are in parallel and have a diameter of ¼ in each. The net area is most nearly

- (A) 2.0 in²
- (B) 2.2 in²
- (C) 2.5 in²
- (D) 2.7 in²

3. A W-shape member (yield strength of 36 ksi; ultimate strength of 58 ksi) carries an axial live tensile load of 420 kips. The member's flanges are bolted to a connection bracket. The shear lag factor for the connection is 0.90. The required net area based on the fracture (rupture) criterion is most nearly

- (A) 12 in²
- (B) 17 in²
- (C) 22 in²
- (D) 27 in²

4. An L4 × 4 × ³⁄₈ angle made from A36 steel is used as a tension member as shown. The angle is connected to a gusset plate with ⁵⁄₈ in diameter bolts. The center-to-center bolt spacing is 3 in. The distance from the centroid of the area connected to the plane of connection (the edge), $\bar{x}$, is 1.13 in.

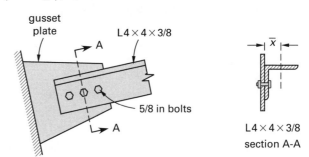

What is most nearly the shear lag factor?

- (A) 0.72
- (B) 0.78
- (C) 0.81
- (D) 0.86

SOLUTIONS

1. For fracture, choose the shortest path, which is along line ABCD, as shown.

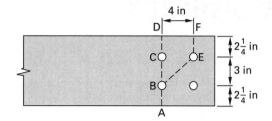

The gross member width is

$$b_g = 2.25 \text{ in} + 3 \text{ in} + 2.25 \text{ in} = 7.5 \text{ in}$$

Calculate the net area.

$$
\begin{aligned}
A_n &= \left(b_g - \sum\left(d_h + \tfrac{1}{16} \text{ in}\right)\right)t \\
&= \left(7.5 \text{ in} - (2)\left(0.75 \text{ in} + \tfrac{1}{16} \text{ in}\right)\right)(0.5 \text{ in}) \\
&= 2.94 \text{ in}^2
\end{aligned}
$$

For bolted members with flat bars, the shear lag factor, U, is 1.0. The effective net area is

$$
\begin{aligned}
A_e &= U A_n = (1.0)(2.94 \text{ in}^2) \\
&= 2.94 \text{ in}^2 \quad (2.9 \text{ in}^2)
\end{aligned}
$$

The answer is (C).

2. The net area excludes area taken up by the holes.

$$
\begin{aligned}
A_n &= \left(b_g - \sum\left(d_h + \tfrac{1}{16} \text{ in}\right)\right)t \\
&= \left(5 \text{ in} - (2 \text{ holes})\left(0.25 \frac{\text{in}}{\text{hole}} + \tfrac{1}{16} \text{ in}\right)\right) \\
&\quad \times (0.5 \text{ in}) \\
&= 2.19 \text{ in}^2 \quad (2.2 \text{ in}^2)
\end{aligned}
$$

The answer is (B).

3. Find the effective net area based on the fracture (rupture) criterion.

$$\phi_t = 0.75$$

$$T_u = \phi_t T_n = \phi_t F_u A_e$$

$$(1.6)(420 \text{ kips}) = (0.75)\left(58 \frac{\text{kips}}{\text{in}^2}\right) A_e$$

$$A_e = 15.45 \text{ in}^2$$

The net area is

$$
\begin{aligned}
A_e &= U A_n \\
A_n &= \frac{A_e}{U} = \frac{15.45 \text{ in}^2}{0.90} \\
&= 17.2 \text{ in}^2 \quad (17 \text{ in}^2)
\end{aligned}
$$

The answer is (B).

4. The length of the connection for bolted connections is the distance between the outer holes.

$$L = (2)(3 \text{ in}) = 6 \text{ in}$$

The shear lag factor is

$$
\begin{aligned}
U &= 1 - \frac{\overline{x}}{L} \\
&= 1 - \frac{1.13 \text{ in}}{6 \text{ in}} \\
&= 0.812 \quad (0.81)
\end{aligned}
$$

The answer is (C).

45 Structural Steel: Beam-Columns

PRACTICE PROBLEMS

1. A W14 × 120, A992 steel beam has been chosen to carry an axial live compressive load of 140 kips and a factored 480 ft-kips live moment about the strong axis. The unsupported length is 20 ft. Sidesway is permitted in the direction of bending. $K = 1.0$. The compressive strength is 780 kips, and the bending strength is 495 ft-kips. The beam-column is subjected to

- (A) small compression and is adequate
- (B) large compression and is adequate
- (C) small compression and is inadequate
- (D) large compression and is inadequate

2. A W14 × 132 beam has been chosen to carry an axial live load of 160 kips and a maximum live moment of 320 ft-kips about the strong axis. The unsupported length is 32 ft. $K = 1$, $C_m = 1$, and $I_x = 1530$ in^4. Taking into account the second-order effects, what is most nearly the required flexural strength?

- (A) 340 ft-kips
- (B) 380 ft-kips
- (C) 420 ft-kips
- (D) 460 ft-kips

SOLUTIONS

1. To determine whether the member is subjected to small or large compression, find the ratio of compressive load to compressive strength.

$$\frac{P_r}{\phi P_n} = \frac{140 \text{ kips}}{780 \text{ kips}} = 0.18$$

Since $P_r/\phi P_n < 0.2$, the member is subjected to a small compression. Determine if the member is adequate.

$$\frac{P_r}{2(\phi P_n)} + \frac{M_r}{\phi M_{nx}} = \frac{140 \text{ kips}}{(2)(780 \text{ kips})} + \frac{480 \text{ ft-kips}}{495 \text{ ft-kips}}$$
$$= 1.06$$

Since the computed value is greater than 1.0, the member is inadequate.

The answer is (C).

2. Find the Euler buckling load.

$$P_{e1} = \frac{\pi^2 EI}{(KL)_x^2} = \frac{\pi^2 \left(29{,}000 \, \frac{\text{kips}}{\text{in}^2}\right)(1530 \text{ in}^4)}{\left((1)(32 \text{ ft})\left(12 \, \frac{\text{in}}{\text{ft}}\right)\right)^2}$$
$$= 2970 \text{ kips}$$

Find the x-x axis flexural magnifier.

$$B_1 = \frac{C_m}{1 - \dfrac{P_r}{P_{e1}}} = \frac{1}{1 - \dfrac{160 \text{ kips}}{2970 \text{ kips}}}$$
$$= 1.057$$

The required flexural strength is

$$M_r = B_1 M_{nt} = (1.057)(320 \text{ ft-kips})$$
$$= 338.2 \text{ ft-kips} \quad (340 \text{ ft-kips})$$

The answer is (A).

Structural Design

PRACTICE PROBLEMS

1. The connection shown consists of 11 grade A307 $^{7}/_{8}$ in diameter bolts. Bolt hole sizes are standard. The ultimate strength of the connected member is 58 ksi. The connected member is 0.5 in thick. The edge clear distance is 2.5 in, and the center-to-center spacing of the holes is 3 in.

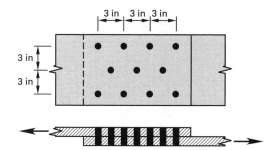

The available bearing strength per bolt per inch of thickness in the connection is most nearly

(A) 76 kips/in

(B) 83 kips/in

(C) 91 kips/in

(D) 110 kips/in

2. A connection is made from two $^{3}/_{4}$ in bolts in parallel, placed on a 8.5 in wide steel bar. The bar is 1 in thick that has an ultimate strength of 65 ksi. The holes are 3 in from their centers to the side of the bar and 2.25 in from center to center. Bolt hole sizes are standard. The bearing resistance of the entire connection is most nearly

(A) 84 kips

(B) 95 kips

(C) 170 kips

(D) 210 kips

SOLUTIONS

1. Determine the bolt hole diameter.

$$d_h = d_b + \tfrac{1}{16} \text{ in}$$
$$= \tfrac{7}{8} \text{ in} + \tfrac{1}{16} \text{ in}$$
$$= \tfrac{15}{16} \text{ in}$$

Determine which clear distance, L_c, value controls for the bolts. Obtain the clear distance in between the interior holes.

$$L_c = s - d_h$$
$$= 3 \text{ in} - \tfrac{15}{16} \text{ in}$$
$$= 2\tfrac{1}{16} \text{ in} < 2\tfrac{1}{2} \text{ in provided}$$

The smaller value for the clear distance controls. Determine the available bearing strength.

$$\phi r_n = \phi 1.2 L_c F_u \le \phi 2.4 d_b F_u$$
$$= (0.75)(1.2)\left(2\tfrac{1}{16} \text{ in}\right)\left(58 \ \frac{\text{kips}}{\text{in}^2}\right)$$
$$\le (0.75)(2.4)\left(\tfrac{7}{8} \text{ in}\right)\left(58 \ \frac{\text{kips}}{\text{in}^2}\right)$$
$$= 107.7 \ \frac{\text{kips}}{\text{in}} \le 91.35 \ \frac{\text{kips}}{\text{in}}$$

Since the available bearing strength determined by the clear distance is larger than the available bearing strength determined by the bolt size, the smaller value controls, and the available bearing strength is 91 kips/in.

This value may also be determined using AISC tables for a center-to-center spacing of 3 in, a connected member strength of 58 ksi, and a bolt size of $^{7}/_{8}$ in.

The answer is (C).

2. Calculate the hole diameter using the bolt diameter and standard clearance.

$$d_h = d_b + \tfrac{1}{16} \text{ in}$$
$$= \tfrac{3}{4} \text{ in} + \tfrac{1}{16} \text{ in}$$
$$= \tfrac{13}{16} \text{ in} \quad (0.8125 \text{ in})$$

The center-to-center spacing of the interior holes is 3 in. Obtain the clear distance between the interior holes.

$$L_c = s - d_h$$
$$= 2.25 \text{ in} - 0.8125 \text{ in}$$
$$= 1.4375 \text{ in}$$

Determine the clear distance between the hole and the edge of the steel bar.

$$L_c = L_e - \frac{d_h}{2}$$
$$= 3 \text{ in} - \frac{0.8125 \text{ in}}{2}$$
$$= 2.594 \text{ in}$$

Use the smaller clear distance. Since each bolt has the same properties, calculate the available bearing strength for one bolt.

$$r_n = \min \begin{cases} \phi 1.2 L_c F_u \\ \phi 2.4 d_b F_u \end{cases}$$

$$= \min \begin{cases} (0.75)(1.2)(1.435 \text{ in}) \left(65 \ \frac{\text{kips}}{\text{in}^2} \right) \\ (0.75)(2.4)(0.75 \text{ in}) \left(65 \ \frac{\text{kips}}{\text{in}^2} \right) \end{cases}$$

$$= \min \begin{cases} 84.09 \ \dfrac{\text{kips}}{\text{in}} \\ 87.75 \ \dfrac{\text{kips}}{\text{in}} \end{cases}$$

$$= 84.09 \text{ kips/in}$$

The available bearing strength is

$$\phi R_n = \sum \phi r_n t$$
$$= (2 \text{ bolts}) \left(84.09 \ \frac{\text{kip}}{\text{bolt-in}} \right) (1 \text{ in})$$
$$= 168.2 \text{ kips} \quad (170 \text{ kips})$$

The answer is (C).

47 Transportation Planning and Capacity

PRACTICE PROBLEMS

1. An interstate weigh station can weigh an average of 20 trucks per hour. Trucks arrive at the average rate of 12 trucks per hour. Performance is described by an $M/M/1$ model. What is most nearly the steady-state value of the trucks' time spent waiting to be weighed?

(A) 0.13 hr/truck

(B) 0.25 hr/truck

(C) 4.0 hr/truck

(D) 8.0 hr/truck

2. A traffic flow relationship is given by $q = kv$, where q is the traffic volume in veh/hr, k is the traffic density in veh/mi, and v is the mean speed in mi/hr. The mean speed on a road in mi/hr is given by the relationship

$$v = 60 \ \frac{mi}{hr} - \left(0.2 \ \frac{mi^2}{veh\text{-}hr}\right) k$$

The maximum capacity of overall traffic volume for this road is most nearly

(A) 3400 veh/hr

(B) 4300 veh/hr

(C) 4500 veh/hr

(D) 5000 veh/hr

3. A highway weigh station can inspect an average of 16 trucks per hour. Trucks arrive at the average rate of 10 trucks per hour. Performance is described by an $M/M/1$ model. What is most nearly the steady-state value of the number of trucks in the station?

(A) 0.16

(B) 0.60

(C) 1.7

(D) 2.5

4. Vehicles arrive at a parking garage at an average rate of 45 vehicles per hour. Vehicles are parked by either of two attendants at an average rate of one vehicle per minute for each attendant. The queue discipline is first come, first served. The arrival of vehicles is modeled as a Poisson's distribution and attendant parking time is modeled as an exponential distribution. What is most nearly the probability that at least one attendant will be idle?

(A) 0.34

(B) 0.45

(C) 0.52

(D) 0.80

5. The relationship between traffic density and mean vehicle speed is shown for a particular road.

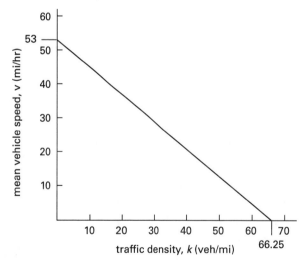

What is most nearly the maximum traffic volume for this road?

(A) 760 veh/hr

(B) 880 veh/hr

(C) 900 veh/hr

(D) 960 veh/hr

6. A border's point of entry can weigh an average of 24 trucks per hour. Trucks arrive at the average rate of 15 trucks per hour. Performance is described by an $M/M/1$ model. What is most nearly the steady-state value of the probability that there will be five trucks being weighed or waiting to be weighed at any time?

(A) 0.010

(B) 0.036

(C) 0.38

(D) 0.63

SOLUTIONS

1. The average time spent waiting to be weighed is

$$W = \frac{1}{\mu - \lambda} = \frac{1}{20\ \dfrac{\text{trucks}}{\text{hr}} - 12\ \dfrac{\text{trucks}}{\text{hr}}}$$

$$= 0.125\ \text{hr/truck} \quad (0.13\ \text{hr/truck})$$

The answer is (A).

2. The mean speed relationship can be substituted into the traffic flow relationship, resulting in a quadratic relationship (i.e., a parabolic curve).

$$q = kv = k\left(60\ \frac{\text{mi}}{\text{hr}} - \left(0.2\ \frac{\text{mi}^2}{\text{veh-hr}}\right)k\right)$$

$$= \left(60\ \frac{\text{mi}}{\text{hr}}\right)k - \left(0.2\ \frac{\text{mi}^2}{\text{veh-hr}}\right)k^2$$

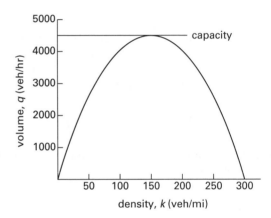

To determine the traffic volume capacity, it is necessary to find the maximum point on the parabolic curve (i.e., the location where the slope of the curve equals 0).

$$\frac{dq}{dk} = 0$$

$$\frac{d\left(\begin{array}{l}\left(60\ \frac{\text{mi}}{\text{hr}}\right)k \\ - \left(0.2\ \frac{\text{mi}^2}{\text{veh-hr}}\right)k^2\end{array}\right)}{dk} = 60\ \frac{\text{mi}}{\text{hr}} - \left(0.4\ \frac{\text{mi}^2}{\text{veh-hr}}\right)k$$

$$= 0$$

$$k = \frac{60\ \dfrac{\text{mi}}{\text{hr}}}{0.4\ \dfrac{\text{mi}^2}{\text{veh-hr}}}$$

$$= 150\ \text{veh/mi}$$

Substituting $k = 150$ veh/mi into the traffic flow relationship gives

$$q = \left(60 \ \frac{\text{mi}}{\text{hr}}\right)k - \left(0.2 \ \frac{\text{mi}^2}{\text{veh-hr}}\right)k^2$$

$$= \left(60 \ \frac{\text{mi}}{\text{hr}}\right)\left(150 \ \frac{\text{veh}}{\text{mi}}\right)$$

$$\quad - \left(0.2 \ \frac{\text{mi}^2}{\text{veh-hr}}\right)\left(150 \ \frac{\text{veh}}{\text{mi}}\right)^2$$

$$= 4500 \ \text{veh/hr}$$

The answer is (C).

3. The average number of trucks in the station is

$$L = \frac{\lambda}{\mu - \lambda} = \frac{10 \ \dfrac{\text{trucks}}{\text{hr}}}{16 \ \dfrac{\text{trucks}}{\text{hr}} - 10 \ \dfrac{\text{trucks}}{\text{hr}}}$$

$$= 1.66 \quad (1.7)$$

The answer is (C).

4. The mean arrival rate of vehicles per minute is

$$\lambda = \frac{45}{(1 \ \text{hr})\left(60 \ \dfrac{\text{min}}{\text{hr}}\right)} = 0.75 \ 1/\text{min}$$

The service rate per server is $\mu = 1 \ 1/\text{min}$.

For two servers ($s = 2$),

$$\rho = \frac{\lambda}{s\mu} = \frac{0.75 \ \dfrac{1}{\text{min}}}{(2)\left(1 \ \dfrac{1}{\text{min}}\right)} = 0.375$$

Next, calculate the probability, P_0, of both attendants being idle (i.e., no units in the system).

$$P_0 = \frac{1 - \rho}{1 + \rho} = \frac{1 - 0.375}{1 + 0.375} = 0.4545$$

Calculate the probability, P_1, of one attendant being idle.

$$P_1 = P_0\left(\frac{\left(\frac{\lambda}{\mu}\right)^n}{n!}\right) = (0.4545)\left(\frac{\left(\dfrac{0.75 \ \frac{1}{\text{min}}}{1 \ \frac{1}{\text{min}}}\right)^1}{1!}\right)$$

$$= 0.3409$$

Finally, add P_0 and P_1 to calculate the probability of at least one idle attendant.

$$P = P_0 + P_1 = 0.4545 + 0.3409$$

$$= 0.7954 \quad (0.80)$$

The answer is (D).

5. From the mean vehicle speed versus traffic density graph, the relationship is linear.

$$v = 53 \ \frac{\text{mi}}{\text{hr}} - \left(\frac{53 \ \dfrac{\text{mi}}{\text{hr}}}{66.25 \ \dfrac{\text{veh}}{\text{mi}}}\right)k$$

$$= 53 \ \frac{\text{mi}}{\text{hr}} - \left(0.8 \ \frac{\text{mi}^2}{\text{veh-hr}}\right)k \quad [\text{velocity in mi/hr}]$$

This expression for speed can be substituted into the standard relationship for traffic volume.

$$q = kv$$

$$= k\left(53 \ \frac{\text{mi}}{\text{hr}} - \left(0.8 \ \frac{\text{mi}^2}{\text{veh-hr}}\right)k\right)$$

$$= \left(53 \ \frac{\text{mi}}{\text{hr}}\right)k - \left(0.8 \ \frac{\text{mi}^2}{\text{veh-hr}}\right)k^2$$

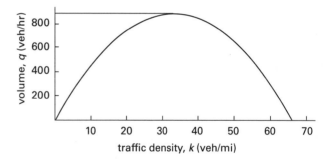

The maximum traffic volume occurs where the slope of this curve equals 0.

$$\frac{dq}{dk} = \frac{d\left(\left(53 \ \frac{\text{mi}}{\text{hr}}\right)k - \left(0.8 \ \frac{\text{mi}^2}{\text{veh-hr}}\right)k^2\right)}{dk}$$

$$= 53 \ \frac{\text{mi}}{\text{hr}} - \left(1.6 \ \frac{\text{mi}^2}{\text{veh-hr}}\right)k$$

$$= 0$$

$$k_{\text{max} \, q} = \frac{53 \ \dfrac{\text{mi}}{\text{hr}}}{1.6 \ \dfrac{\text{mi}^2}{\text{veh-hr}}}$$

$$= 33.125 \ \text{veh/mi}$$

The density is 33.125 veh/mi at maximum traffic volume.

$$q = \left(53 \ \frac{\text{mi}}{\text{hr}}\right) k - \left(0.8 \ \frac{\text{mi}^2}{\text{veh-hr}}\right) k^2$$

$$= \left(53 \ \frac{\text{mi}}{\text{hr}}\right)\left(33.125 \ \frac{\text{veh}}{\text{mi}}\right)$$

$$\quad - \left(0.8 \ \frac{\text{mi}^2}{\text{veh-hr}}\right)\left(33.125 \ \frac{\text{veh}}{\text{mi}}\right)^2$$

$$= 878 \ \text{veh/hr} \quad (880 \ \text{veh/hr})$$

The answer is (B).

6. The utilization factor is

$$\rho = \frac{\lambda}{s\mu} = \frac{15 \ \dfrac{\text{trucks}}{\text{hr}}}{(1)\left(24 \ \dfrac{\text{trucks}}{\text{hr}}\right)} = 0.625$$

The probability of five trucks being weighed or waiting to be weighed is

$$P_0 = 1 - \rho = 1 - 0.625 = 0.375$$
$$P_5 = P_0 \rho^n = (0.375)(0.625)^5 = 0.036$$

The answer is (B).

48 Plane Surveying

PRACTICE PROBLEMS

1. Boundary and traverse lines bounding an irregular area are shown.

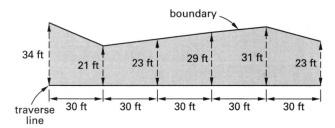

The total area between the irregular boundary and the traverse line is most nearly

(A) 3600 ft^2

(B) 3800 ft^2

(C) 4000 ft^2

(D) 4200 ft^2

2. Global positioning system (GPS) latitudes and longitudes were taken of a plot of land. In the region where the plot is located, the length of a degree of latitude is 364,320 ft, and the length of a degree of longitude is 248,160 ft.

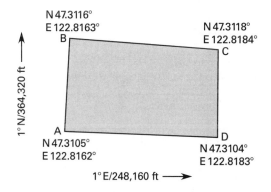

What is most nearly the area of the plot?

(A) 5.0 ac

(B) 5.1 ac

(C) 5.3 ac

(D) 5.4 ac

3. The illustration shows a curve from $x = 0$ to $x = 6$.

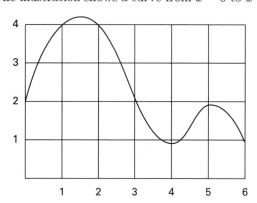

Using intervals of 1, what is most nearly the area under the curve predicted by Simpson's $^1/_3$ rule?

(A) 14

(B) 15

(C) 16

(D) 17

4. A polygon is created by enclosing lines as shown.

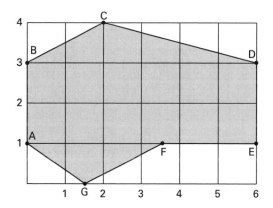

Using an interval of 1, what is most nearly the area of the polygon predicted by the trapezoidal rule?

(A) 15

(B) 16

(C) 17

(D) 18

5. Which statement(s) concerning methods used to determine areas under a curve or line is/are true?

I.   The trapezoidal rule applies to areas where the irregular sides are curved.

II. Simpson's rule only applies to an odd number of data points.

III. The area by coordinates method can be used if the coordinates of the traverse leg end points are known.

  (A) I only

  (B) III only

  (C) I and III

  (D) I, II, and III

SOLUTIONS

1. The trapezoidal rule is

$$\text{area} = w\left(\frac{h_1 + h_6}{2} + h_2 + h_3 + h_4 + h_5\right)$$

$$= (30 \text{ ft})\left(\begin{array}{c}\dfrac{34 \text{ ft} + 23 \text{ ft}}{2} + 21 \text{ ft} \\ + 23 \text{ ft} + 29 \text{ ft} + 31 \text{ ft}\end{array}\right)$$

$$= 3975 \text{ ft}^2 \quad (4000 \text{ ft}^2)$$

The answer is (C).

2. Use the area by coordinates method to find the area of the plot.

$$\text{area} = \frac{\begin{array}{c}X_A(Y_B - Y_D) + X_B(Y_C - Y_A) \\ + X_C(Y_D - Y_B) \\ + X_D(Y_A - Y_C)\end{array}}{2}$$

$$= \frac{\begin{array}{c}(122.8162°)(47.3116° - 47.3104°) \\ + (122.8163°)(47.3118° - 47.3105°) \\ + (122.8184°)(47.3104° - 47.3116°) \\ + (122.8183°)(47.3105° - 47.3118°)\end{array}}{2}$$

$$= -2.62 \times 10^{-6} \text{ deg}^2$$

Use the lengths per degree of latitude and longitude to convert the angles to distance.

$$\text{area} = \frac{(-2.62 \times 10^{-6} \text{ deg}^2)\left(248{,}160 \, \dfrac{\text{ft}}{°\text{E}}\right)\left(364{,}320 \, \dfrac{\text{ft}}{°\text{N}}\right)}{43{,}560 \, \dfrac{\text{ft}^2}{\text{ac}}}$$

$$= 5.44 \text{ ac} \quad (5.4 \text{ ac})$$

The answer is (D).

3. There is an even number of intervals, so Simpson's rule can be used. Use the graph to determine the measurement at each value of x. The starting measurement is at $x = 0$. That is, $h_1 = 2$.

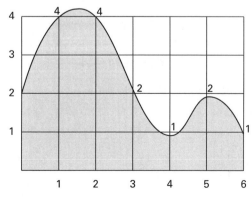

The area under the curve using Simpson's $\frac{1}{3}$ rule is

$$\text{area} = \frac{w\left(h_1 + 2\left(\sum_{k=3,5,\ldots}^{n-2} h_k\right) + 4\left(\sum_{k=2,4,\ldots}^{n-1} h_k\right) + h_n\right)}{3}$$

$$= \frac{(1)\left(2 + (2)(4+1) + (4)(4+2+2) + 1\right)}{3}$$

$$= 15$$

The answer is (B).

4. The area of the polygon can be determined by subtracting the area under the bottom line from the area under the top line.

Since the common interval is 1, there will be seven height measurements (i.e., $n = 7$). The height of each trapezoid can be determined as shown in the figure.

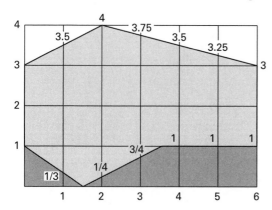

The area under the top line is

$$\text{area}_{\text{top}} = w\left(\frac{h_1 + h_7}{2} + h_2 + h_3 + h_4 + h_5 + h_6\right)$$

$$= (1)\left(\frac{3+3}{2} + 3.5 + 4 + 3.75 + 3.5 + 3.25\right)$$

$$= 21$$

The area under the bottom line is

$$\text{area}_{\text{bottom}} = w\left(\frac{h_1 + h_7}{2} + h_2 + h_3 + h_4 + h_5 + h_6\right)$$

$$= (1)\left(\frac{1+1}{2} + \frac{1}{3} + \frac{1}{4} + \frac{3}{4} + 1 + 1\right)$$

$$= 4.333$$

The area of the polygon is

$$\text{area} = \text{area}_{\text{top}} - \text{area}_{\text{bottom}}$$

$$= 21 - 4.333$$

$$= 16.667 \quad (17)$$

The answer is (C).

5. The trapezoidal rule is applicable to areas where the irregular sides are straight or nearly straight. Option I is false. Simpson's rule is applicable to areas whose irregular sides are curved, but applies only to an even number of points. Option II is false. The area by coordinates method calculates the traverse area using the coordinates of the leg end points. Option III is true.

The answer is (B).

Geometric Design

PRACTICE PROBLEMS

1. A horizontal curve is laid out with the point of curve, PC, station and the length of long chord, LC, as shown.

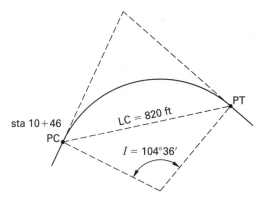

The radius of the curve is most nearly

(A) 520 ft

(B) 560 ft

(C) 620 ft

(D) 670 ft

2. A freeway route has a horizontal curve with a PI at sta 11+01.86, an intersection angle, I, of $12°24'00''$ right, and a radius of 1760 ft. The PC is located at

(A) sta 9+10

(B) sta 9+22

(C) sta 10+11

(D) sta 10+24

3. A vertical sag curve has a length of 8 sta and connects a −2.0% grade to a 1.6% vertical grade. The PVI is located at sta 87+00 and has an elevation of 2438 ft.

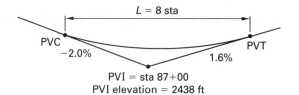

The elevation of the lowest point on the vertical curve is most nearly

(A) 2420 ft

(B) 2430 ft

(C) 2440 ft

(D) 2450 ft

4. A 6° curve has forward and back tangents that intersect at sta 14+87.33.

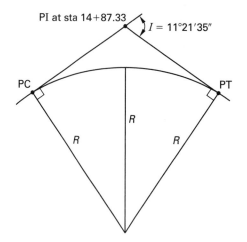

The station of the point of curve, PC, is most nearly

(A) sta 5+32

(B) sta 9+93

(C) sta 11+28

(D) sta 13+92

5. A horizontal curve is laid out with the point of curve, PC, station and the length of long chord, LC, as shown.

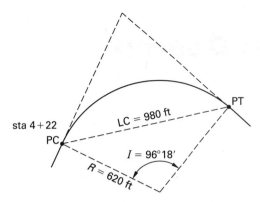

The curve radius, R, is 620 ft. With stationing around the curve, the stationing of the point of tangent, PT, is most near to

(A) sta 6+20

(B) sta 10+42

(C) sta 14+02

(D) sta 14+64

SOLUTIONS

1. The intersection angle, I, in decimal degrees is

$$I = 104° + \frac{36 \text{ min}}{60 \frac{\text{min}}{\text{deg}}} = 104.6°$$

The radius of the curve is

$$R = \frac{\text{LC}}{2 \sin \frac{I}{2}} = \frac{820 \text{ ft}}{2 \sin \frac{104.6°}{2}}$$

$$= 518.18 \text{ ft} \quad (520 \text{ ft})$$

The answer is (A).

2. Convert the intersection angle to a decimal value.

$$I = 12° + \frac{24 \text{ min}}{60 \frac{\text{min}}{\text{deg}}} = 12.4°$$

The tangent length is

$$T = R \tan \frac{I}{2}$$

$$= (1760 \text{ ft}) \tan \left(\frac{12.4°}{2} \right)$$

$$= 191.20 \text{ ft}$$

The PC is located at

$$\text{sta PC} = \text{sta PI} - T$$

$$= 1101.86 \text{ ft} - 191.20 \text{ ft}$$

$$= 910 \text{ ft} \quad (\text{sta } 9+10)$$

The answer is (A).

3. The PVI is located at the curve's midpoint. The elevation of the PVC is

$$Y_{\text{PVC}} = Y_{\text{PVI}} + |g_1| \frac{L}{2}$$

$$= 2438 \text{ ft} + (0.02) \left(\frac{8 \text{ sta}}{2} \right) \left(100 \frac{\text{ft}}{\text{sta}} \right)$$

$$= 2446 \text{ ft}$$

The distance from the PVC to the lowest point on the curve is

$$x_m = \frac{g_1 L}{g_1 - g_2}$$

$$= \frac{(-0.02)(8 \text{ sta}) \left(100 \frac{\text{ft}}{\text{sta}} \right)}{-0.02 - 0.016}$$

$$= 444.44 \text{ ft}$$

Determine the elevation at the lowest point.

$$Y = Y_{\text{PVC}} + g_1 x + \left(\frac{g_2 - g_1}{2L}\right) x^2$$

$$= 2446 \text{ ft} + (-0.02)(444.44 \text{ ft})$$

$$+ \left(\frac{0.016 - (-0.02)}{(2)(8 \text{ sta})\left(100 \dfrac{\text{ft}}{\text{sta}}\right)}\right)(444.44 \text{ ft})^2$$

$$= 2441.55 \text{ ft} \quad (2440 \text{ ft})$$

The answer is (C).

4. The radius of the curve is

$$R = \frac{5729.58}{D} = \frac{5729.58 \text{ ft-deg}}{6°}$$

$$= 954.93 \text{ ft}$$

Convert the intersection angle to a decimal value.

$$I = 11° + \frac{21 \text{ min}}{60 \dfrac{\text{min}}{\text{deg}}} + \frac{35 \text{ sec}}{\left(60 \dfrac{\text{sec}}{\text{min}}\right)\left(60 \dfrac{\text{min}}{\text{deg}}\right)}$$

$$= 11.36°$$

The tangent length is

$$T = R \tan \frac{I}{2} = (954.93 \text{ ft}) \tan\left(\frac{11.36°}{2}\right)$$

$$= 94.98 \text{ ft}$$

The station of the point of curve, PC, is

$$\text{sta PC} = \text{sta PI} - T$$

$$= 1487.33 \text{ ft} - 94.98 \text{ ft}$$

$$= 1392 \text{ ft} \quad (\text{sta } 13{+}92)$$

The answer is (D).

5. The intersection angle, I, in decimal degrees is

$$I = 96° + \frac{18 \text{ min}}{60 \dfrac{\text{min}}{\text{deg}}} = 96.3°$$

The length of curve from PC to PT is

$$L = RI\left(\frac{\pi}{180°}\right)$$

$$= (620 \text{ ft})(96.3°)\left(\frac{\pi \text{ rad}}{180°}\right)$$

$$= 1042.07 \text{ ft}$$

The stationing of the point of tangent is

$$\text{sta PT} = \text{sta PC} + L$$

$$= 422 \text{ ft} + 1042.07 \text{ ft}$$

$$= 1464 \text{ ft} \quad (\text{sta } 14{+}64)$$

The answer is (D).

Transportation/
Surveying

50 Earthwork

PRACTICE PROBLEMS

1. Earthwork quantities for a section of roadway indicate a transition from fill to cut. The following areas are scaled from the print cross sections. Where there are transitions between cut and fill, the cut and fill roadway cross sections are both triangular in shape.

station	cut (ft^2)	fill (ft^2)
20+00	–	1864.42
20+10.50	–	468.88
20+21.50	154.14	103.66
20+28.45	696.75	–
20+40	2017.37	–

The total volume of fill required for this section of road is most nearly

(A) 11,000 ft^3

(B) 16,000 ft^3

(C) 19,000 ft^3

(D) 21,000 ft^3

2. Consider the borrow pit grid shown. Existing excavation depths (in feet) are shown for each corner.

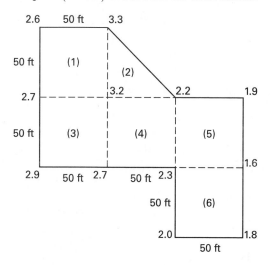

The total undercut volume of this borrow pit is most nearly

(A) 1300 yd^3

(B) 1600 yd^3

(C) 1900 yd^3

(D) 2100 yd^3

3. The table represents the areas of cut and fill at each roadway station along a rural road project.

station	cut (ft^2)	fill (ft^2)
10+00	1600	270
20+00	0	810
30+00	1100	270

The amount of borrow or waste between the stations is most nearly

(A) 23,000 ft^3 borrow

(B) 24,000 ft^3 waste

(C) 25,000 ft^3 borrow

(D) 27,000 ft^3 waste

4. Earthwork quantities for a section of roadway indicate a transition from fill to cut. The following areas are scaled from the print cross sections. Where there are transitions between cut and fill, the cut and fill roadway cross sections are both triangular in shape.

station	cut (ft^2)	fill (ft^2)
10+30	–	126.5
10+60	160.7	50.6
10+82	505.0	–

The total volume of cut required for this section of road is most nearly

(A) 3030 ft^3

(B) 4900 ft^3

(C) 7200 ft^3

(D) 9700 ft^3

SOLUTIONS

1. Earthwork volumes for fill areas and cut areas can be calculated using the average end area formula. Since the cut and fill areas are triangular in shape, earthwork volumes in the transition region from fill to cut can be calculated from the formula that gives the volume of a pyramid.

For sta 20+00 to sta 20+10.50,

$$L = 10.50 \text{ ft} - 0 \text{ ft} = 10.50 \text{ ft}$$

The fill volume is

$$V_{\text{fill}} = L\left(\frac{A_1 + A_2}{2}\right)$$
$$= (10.50 \text{ ft})\left(\frac{1864.42 \text{ ft}^2 + 468.88 \text{ ft}^2}{2}\right)$$
$$= 12{,}249.83 \text{ ft}^3$$

For sta 20+10.50 to sta 20+21.50,

$$L = 21.50 \text{ ft} - 10.50 \text{ ft} = 11.00 \text{ ft}$$

The fill volume is

$$V_{\text{fill}} = L\left(\frac{A_1 + A_2}{2}\right)$$
$$= (11.00 \text{ ft})\left(\frac{468.88 \text{ ft}^2 + 103.66 \text{ ft}^2}{2}\right)$$
$$= 3148.97 \text{ ft}^3$$

Cut is required at sta 20+21.50, but no cut is required at sta 20+10.50, so a cut-to-fill transition occurs between these two points. (The cut area could conceivably decrease to zero at any point before sta 20+21.50, but a reasonable assumption is that data was given at sta 20+10.50 for a reason: that is the point where the cut becomes zero.)

The cut area cross section is triangular at sta 20+21.50, tapering to zero at sta 20+10.50, so the soil mass is essentially a triangular-base pyramid on its side, with the apex at sta 20+10.50. The "height" of this pyramid is 11.00 ft. Calculate the volume of a pyramid of cut.

$$V_{\text{cut}} = h\left(\frac{\text{area of base}}{3}\right)$$
$$= (11.00 \text{ ft})\left(\frac{154.14 \text{ ft}^2}{3}\right)$$
$$= 565.18 \text{ ft}^3$$

For sta 20+21.50 to sta 20+28.45,

$$L = 28.45 \text{ ft} - 21.50 \text{ ft}$$
$$= 6.95 \text{ ft}$$

Fill is required at sta 20+21.50, but no fill is required at sta 20+28.45, so a fill-to-cut transition occurs between these two points. The fill area cross section is triangular at sta 20+21.50, tapering to zero at sta 20+28.45. The "height" of the pyramid is 6.95 ft. Calculate the volume of a pyramid of fill.

$$V_{\text{fill}} = h\left(\frac{\text{area of base}}{3}\right)$$
$$= (6.95 \text{ ft})\left(\frac{103.66 \text{ ft}^2}{3}\right)$$
$$= 240.15 \text{ ft}^3$$

The cut volume is

$$V_{\text{cut}} = L\left(\frac{A_1 + A_2}{2}\right)$$
$$= (6.95 \text{ ft})\left(\frac{154.14 \text{ ft}^2 + 696.75 \text{ ft}^2}{2}\right)$$
$$= 2956.84 \text{ ft}^3$$

For sta 20+28.45 to sta 20+40,

$$L = 40 \text{ ft} - 28.45 \text{ ft}$$
$$= 11.55 \text{ ft}$$

The cut volume is

$$V_{\text{cut}} = L\left(\frac{A_1 + A_2}{2}\right)$$
$$= (11.55 \text{ ft})\left(\frac{696.75 \text{ ft}^2 + 2017.37 \text{ ft}^2}{2}\right)$$
$$= 15{,}674.04 \text{ ft}^3$$

A table that summarizes earthwork volumes is now made.

station	cut area (ft^2)	fill area (ft^2)	cut volume (ft^3)	fill volume (ft^3)
20+00	–	1864.42		
			–	12,249.83
20+10.50	–	468.88		
			565.18	3148.97
20+21.50	154.14	103.66		
			2956.84	240.15
20+28.45	696.75	–		
			15,674.04	–
20+40	2017.37	–		
	total		19,196.06	15,638.95

Therefore, the total volume of fill required for this section of road is 15,638.95 ft^3 (16,000 ft^3).

The answer is (B).

2. Calculate the average depth of undercut by summing the undercut depths at each of the corners and dividing the total by the number of corners. Calculate the undercut volume by multiplying the area by the average depth of undercut.

The area of each full cell is

$$(50 \text{ ft})(50 \text{ ft}) = 2500 \text{ ft}^2$$

The area of each triangular half-cell is

$$\frac{2500 \text{ ft}^2}{2} = 1250 \text{ ft}^2$$

cell number	area (ft^2)	average depth of undercut (ft)	volume (ft^3)
1	2500	2.95	7375
2	1250	2.90	3625
3	2500	2.88	7200
4	2500	2.60	6500
5	2500	2.00	5000
6	2500	1.93	4825
		total	34,525

The total volume is

$$V_{\text{total}} = \frac{34{,}525 \text{ ft}^3}{\left(3 \dfrac{\text{ft}}{\text{yd}}\right)^3}$$

$$= 1279 \text{ yd}^3 \quad (1300 \text{ yd}^3)$$

The answer is (A).

3. Use the average end area method. 100 ft stations are used. The cut volumes are

$$V_{\text{cut,sta 1 to 2}} = L\left(\frac{A_1 + A_2}{2}\right)$$

$$= (100 \text{ ft})\left(\frac{1600 \text{ ft}^2 + 0 \text{ ft}^2}{2}\right)$$

$$= 80{,}000 \text{ ft}^3$$

$$V_{\text{cut,sta 2 to 3}} = L\left(\frac{A_1 + A_2}{2}\right)$$

$$= (100 \text{ ft})\left(\frac{0 \text{ ft}^2 + 1100 \text{ ft}^2}{2}\right)$$

$$= 55{,}000 \text{ ft}^3$$

The fill volumes are

$$V_{\text{fill,sta 1 to 2}} = L\left(\frac{A_1 + A_2}{2}\right)$$

$$= (100 \text{ ft})\left(\frac{270 \text{ ft}^2 + 810 \text{ ft}^2}{2}\right)$$

$$= 54{,}000 \text{ ft}^3$$

$$V_{\text{fill,sta 2 to 3}} = L\left(\frac{A_1 + A_2}{2}\right)$$

$$= (100 \text{ ft})\left(\frac{810 \text{ ft}^2 + 270 \text{ ft}^2}{2}\right)$$

$$= 54{,}000 \text{ ft}^3$$

Determine the total cut and fill volumes.

$$\text{cut:} \quad V = 80{,}000 \text{ ft}^3 + 55{,}000 \text{ ft}^3$$

$$= 135{,}000 \text{ ft}^3$$

$$\text{fill:} \quad V = 54{,}000 \text{ ft}^3 + 54{,}000 \text{ ft}^3$$

$$= 108{,}000 \text{ ft}^3$$

There is $135{,}000 \text{ ft}^3 - 108{,}000 \text{ ft}^3 = 27{,}000 \text{ ft}^3$ more waste than borrow.

The answer is (D).

4. Earthwork volumes for fill areas and cut areas can be calculated using the average end area formula. Since the cut and fill areas are triangular in shape, earthwork volumes in the transition region from fill to cut can be calculated from the formula that gives the volume of a pyramid.

For sta 10+30 to sta 10+60,

$$L = 60 \text{ ft} - 30 \text{ ft} = 30 \text{ ft}$$

The fill volume is

$$V_{\text{fill}} = L\left(\frac{A_1 + A_2}{2}\right)$$

$$= (30 \text{ ft})\left(\frac{126.5 \text{ ft}^2 + 50.6 \text{ ft}^2}{2}\right)$$

$$= 2656.5 \text{ ft}^3$$

Cut is required at sta 10+60, but no cut is required at sta 10+30, so a cut-to-fill transition occurs between these two points. (The cut area could conceivably decrease to zero at a point before sta 10+60, but a reasonable assumption is that data was given at sta 10+30 for a reason: that is the point where the cut becomes zero.)

Transportation/ Surveying

The cut area cross section is triangular at sta 10+60, tapering to zero at sta 10+30, so the soil mass is essentially a triangular-base pyramid on its side, with the apex at sta 10+30. The height of this pyramid is 30 ft. Calculate the volume of a pyramid of cut.

$$V_{cut} = h\left(\frac{\text{area of base}}{3}\right)$$
$$= (30 \text{ ft})\left(\frac{160.7 \text{ ft}^2}{3}\right)$$
$$= 1607.0 \text{ ft}^3$$

For sta 10+60 to sta 10+82,

$$L = 82 \text{ ft} - 60 \text{ ft} = 22 \text{ ft}$$

Fill is required at sta 10+60, but is not required at sta 10+82, so a fill-to-cut transition occurs between these two points. The fill area cross section is triangular at sta 10+60, tapering to zero at sta 10+82. The height of the pyramid is 22 ft. Calculate the volume of a pyramid of fill.

$$V_{fill} = h\left(\frac{\text{area of base}}{3}\right)$$
$$= (22 \text{ ft})\left(\frac{50.6 \text{ ft}^2}{3}\right)$$
$$= 371.1 \text{ ft}^3$$

The cut volume is

$$V_{cut} = L\left(\frac{A_1 + A_2}{2}\right)$$
$$= (22 \text{ ft})\left(\frac{160.7 \text{ ft}^2 + 505.0 \text{ ft}^2}{2}\right)$$
$$= 7322.7 \text{ ft}^3$$

A table that summarizes earthwork volumes is now made.

station	cut area (ft^2)	fill area (ft^2)	cut volume (ft^2)	fill volume (ft^3)
10+30	–	126.5	–	2656.5
10+60	160.7	50.6	1607.0	371.1
10+82	505.0	–	7322.7	–
total			8929.7	3027.6

Therefore, the total volume of cut required for this section of road is 3027.6 ft^3 (3030 ft^3).

The answer is (A).

Transportation/Surveying

51 Pavement Design

PRACTICE PROBLEMS

1. A Superpave design mixture for a highway with equivalent single-axle loads (ESALs) $< 10^7$ has a nominal maximum aggregate size of 19 mm. The mixture has been tested and has the following characteristics:

$$\text{air voids} = 4.0\%$$

$$\text{VMA} = 13.2\%$$

$$\text{VFA} = 70\%$$

$$\text{dust-to-binder ratio} = 0.97$$

$$\text{at } N_{\text{int}} = 8 \text{ gyrations}, G_{\text{mm}} = 87.1\%$$

$$\text{at } N_{\text{max}} = 174 \text{ gyrations}, G_{\text{mm}} = 97.5\%$$

Do these characteristics satisfy their corresponding Superpave requirements?

(A) Yes, all the parameters are within an acceptable range.

(B) No, the VMA is excessive.

(C) No, the dust-to-binder ratio is too high.

(D) No, G_{mm} at N_{max} is too high.

2. What is the approximate load equivalency factor (LEF) for a 32,000 pound tandem-axle truck?

(A) 0.36

(B) 0.86

(C) 7.0

(D) 21

3. A road leading to a stone quarry is traveled by 40 trucks, with each truck making an average of 10 trips per day. When fully loaded, each truck consists of a front single axle transmitting a force of 10,000 lbf and two rear tandem axles, each tandem axle transmitting a force of 20,000 lbf. The load equivalency factor (LEF) for the front single axle is 0.0877. The load equivalency factor for each rear tandem axle is 0.1206.

The 18,000 lbf equivalent single-axle load (ESAL) for the truck traffic on this road for 5 years is most nearly

(A) 0.33 ESALs

(B) 130 ESALs

(C) 48,000 ESALs

(D) 240,000 ESALs

4. A highway pavement design has the material specifications shown.

layer	material	layer coefficient	layer thickness
subbase	sandy gravel	0.11 in^{-1}	12 in
base	crushed stone	0.14 in^{-1}	15 in
surface	asphalt concrete	0.44 in^{-1}	6 in

What is most nearly the structural number of the pavement?

(A) 2

(B) 4

(C) 6

(D) 9

SOLUTIONS

1. This mixture would be designated as a 19 mm Superpave mixture. The limits for such a mixture are

$$\text{air voids} = 4.0\%$$

$$\text{minimum VMA} = 13\%$$

$$\text{VFA} = 65\text{--}75\%$$

$$\text{dust-to-binder ratio} = 0.6\text{--}1.2$$

$$\text{at } N_{\text{int}} = 8 \text{ gyrations, maximum } G_{\text{mm}} = 89\%$$

$$\text{at } N_{\text{max}} = 174 \text{ gyrations, maximum } G_{\text{mm}} = 98\%$$

All parameters in this mixture are within Superpave specifications.

The answer is (A).

2. From a load equivalency factor (LEF) table, the load equivalency factor for a 32,000 pound tandem-axle load is 0.857 (0.86).

The answer is (B).

3. The total ESALs per truck for each trip is

$$\begin{aligned}
\text{ESALs}_{\text{truck}} &= (\text{no. of axles})(\text{LEF}) \\
&= (1 \text{ single axle})(0.0877) \\
&\quad + (2 \text{ tandem axles})(0.1206) \\
&= 0.3289 \text{ ESALs/truck-trip}
\end{aligned}$$

The total daily ESALs for 40 trucks, each making 10 trips a day, is

$$\begin{aligned}
\text{ESALs}_{\text{day}} &= (40 \text{ trucks})\left(10 \ \frac{\text{trips}}{\text{day}}\right)\left(0.3289 \ \frac{\text{ESALs}}{\text{truck-trip}}\right) \\
&= 131.56 \text{ ESALs/day}
\end{aligned}$$

For 5 years, the total ESALs is

$$\begin{aligned}
\text{ESALs}_{5\,\text{yr}} &= (5 \ \text{yr})\left(365 \ \frac{\text{days}}{\text{yr}}\right)\left(131.56 \ \frac{\text{ESALs}}{\text{day}}\right) \\
&= 240{,}097 \text{ ESALs} \quad (240{,}000 \text{ ESALs})
\end{aligned}$$

The answer is (D).

4. The structural number is

$$\begin{aligned}
SN &= a_1 D_1 + a_2 D_2 + a_3 D_3 \\
&= \left(0.44 \ \frac{1}{\text{in}}\right)(6 \text{ in}) + \left(0.14 \ \frac{1}{\text{in}}\right)(15 \text{ in}) \\
&\quad + \left(0.11 \ \frac{1}{\text{in}}\right)(12 \text{ in}) \\
&= 6.06 \quad (6)
\end{aligned}$$

The answer is (C).

52 Traffic Safety

PRACTICE PROBLEMS

1. What is the design perception-reaction time recommended by the American Association of State Highway and Transportation Officials (AASHTO) for calculating stopping sight distance?

   (A) 1.0 sec

   (B) 1.5 sec

   (C) 2.5 sec

   (D) 4.0 sec

2. What is the accepted normal limit of peripheral vision?

   (A) 45°

   (B) 80°

   (C) 160°

   (D) 180°

3. What is the maximum light-intensity-contrast ratio perceptible to the human eye?

   (A) 4:1

   (B) 3:1

   (C) 2:1

   (D) 1:1

4. What percentage of roadway fatalities occur in a work zone annually?

   (A) 1%

   (B) 2%

   (C) 5%

   (D) 10%

SOLUTIONS

1. The typical range of reaction times is 1.5–3 sec. AASHTO recommends a design perception-reaction time of 2.5 sec.

The answer is (C).

2. The accepted normal limit of peripheral vision is 160°, although some individuals have peripheral vision up to 180°.

The answer is (C).

3. The maximum light-intensity-contrast ratio discernible to the human eye is 3:1.

The answer is (B).

4. Work zones account for approximately 2% of all roadway fatalities each year.

The answer is (B).

53 Construction Management, Scheduling, and Estimating

PRACTICE PROBLEMS

1. An activity-on-node diagram for a construction project is given. (Activity letters and durations are shown in each node circle.)

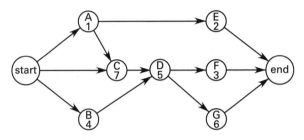

What is most nearly the float time for activity G?

- (A) 0 days
- (B) 1 day
- (C) 6 days
- (D) 25 days

2. For which of the following is the Program Evaluation and Review Technique (PERT) NOT used?

- (A) construction projects
- (B) computer programming assignments
- (C) preparation of bids and proposals
- (D) queueing problems

SOLUTIONS

1. Solve this problem using the critical path method (CPM). There are several paths from the start to the end of this project. Identify the paths and calculate their durations.

start-A-E-end:

$$d = 1 \text{ day} + 2 \text{ days} = 3 \text{ days}$$

start-A-C-D-F-end:

$$d = 1 \text{ day} + 7 \text{ days} + 5 \text{ days} + 3 \text{ days} = 16 \text{ days}$$

start-A-C-D-G-end:

$$d = 1 \text{ day} + 7 \text{ days} + 5 \text{ days} + 6 \text{ days} = 19 \text{ days}$$

start-B-D-F-end:

$$d = 4 \text{ days} + 5 \text{ days} + 3 \text{ days} = 12 \text{ days}$$

start-B-D-G-end:

$$d = 4 \text{ days} + 5 \text{ days} + 6 \text{ days} = 15 \text{ days}$$

The longest path is start-A-C-D-G-end, so this is the critical path. Because activity G is along the critical path, the float time for this activity is 0 days.

The answer is (A).

2. PERT is used to monitor the progress and predict the completion time for large projects. All of the choices given except option (D) are activities that have a finite completion time.

The answer is (D).

54 Procurement and Project Delivery Methods

PRACTICE PROBLEMS

1. Which of the following duties would normally NOT be a responsibility of the estimating department within a general contractor's organization?

    (A) obtaining bid documents

    (B) securing subcontractor/material quotations

    (C) project cost accounting

    (D) delivering competitive or negotiated proposals

2. In competitive bidding, when the bids are all opened, the owner will normally award the contract to the lowest

    (A) available bidder

    (B) qualified bidder

    (C) responsible bidder

    (D) bonded bidder

3. The term "design-build" means the

    (A) design firm designs the project and the client builds it

    (B) design firm both designs and builds the project

    (C) client designs the project and the contractor builds it

    (D) contractor designs the project and the subcontractors build it

SOLUTIONS

1. The estimation process does not include tracking costs or recording actual expenses.

The answer is (C).

2. In open competitive bidding by public and private owners, the bid will be awarded to the lowest responsible bidder.

The lowest responsible bidder is the lowest bidder whose offer best responds in quality, fitness, and capacity to fulfill the particular requirements of the proposed project, and who can fulfill these requirements with the qualifications needed to complete the job in accordance with the terms of the contract.

The answer is (C).

3. *Design-build* is a process where the client interacts only with a single entity, the "design-builder." While the design-builder is usually the general contractor, it can also be the design firm or a partnership consisting of the design firm and contractor.

The answer is (B).

55 Construction Documents

PRACTICE PROBLEMS

1. An activity is coded as 0793–31 62 13.16 using the Construction Specifications Institute's MasterFormat 2012. This coding represents

(A) clearing and grubbing land

(B) demolition of an existing structure

(C) shoring an excavation

(D) driving a deep foundation pile

2. Which of the following organizations is NOT a contributor to the standard design and construction contract documents developed by the Engineers Joint Contract Documents Committee (EJCDC)?

(A) National Society of Professional Engineers

(B) Construction Specifications Institute

(C) Associated General Contractors of America

(D) American Institute of Architects

SOLUTIONS

1. The activity designation is translated as follows.

    project 793
    level 1 activity category 31
    level 2 activity category 62
    level 3 activity category 13

In the MasterFormat specifications, level 1 category 31 is earthwork, level 2 category 62 is driven piles, and level 3 category 13 is concrete piles.

The answer is (D).

2. The Engineers Joint Contract Documents Committee (EJCDC) consists of the National Society of Professional Engineers, the American Council of Engineering Companies (formerly the American Consulting Engineers Council), the American Society of Civil Engineers, Construction Specifications Institute, and the Associated General Contractors of America. The American Institute of Architects is not a member, and it has its own standardized contract documents.

The answer is (D).

Construction

56 Construction Operations and Management

PRACTICE PROBLEMS

1. A crane chart for a crane set up to lift a rooftop onto a 60 ft tall building is shown. The crane's centerline of rotation is 50 ft away from the building, and a 140 ft boom is used at an angle of 60°. The crane is not allowed within 15 ft vertically of the top of the building. What is most nearly the maximum operating radius from the centerline of rotation?

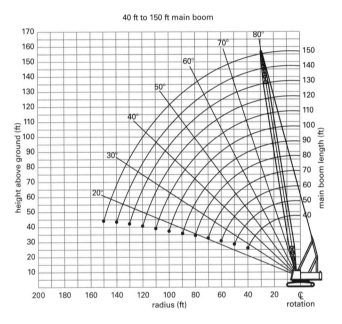

40 ft to 150 ft main boom

(A) 60 ft

(B) 70 ft

(C) 85 ft

(D) 100 ft

2. What factor is responsible for an increase in production rates as a new task is repeated and practiced?

(A) the learning curve

(B) acceleration

(C) phasing

(D) speed enhancement

SOLUTIONS

1. Draw the edges of the building on the chart, then follow the 140 ft boom curve to where it intersects with the 60° pitch line. Draw a vertical line from this intersection to the bottom of the chart. Check for vertical spacing requirements between the roof of the building and the boom of the crane.

The minimum distance from the roof of the building to the boom is 25 ft, which is greater than the 15 ft required. Therefore, the maximum operating radius is approximately 72 ft (70 ft).

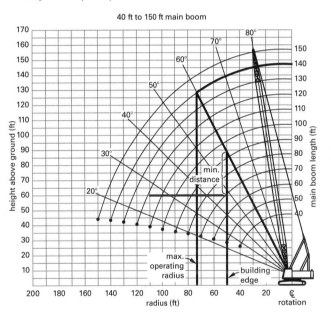

40 ft to 150 ft main boom

The answer is (B).

2. The learning curve is responsible for the usual increase in production rates as workers become skilled at a given task.

The answer is (A).

57 Construction Safety

PRACTICE PROBLEMS

1. What does the crane hand signal shown mean?

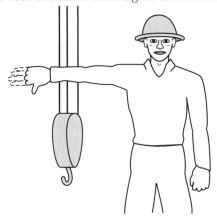

(A) lower the boom

(B) lower the load

(C) raise the boom and lower the load

(D) lower the boom and raise the load

2. In addition to protective goggles, what personal protective equipment is required when welding pipe in a deep trench?

(A) fall protection harness

(B) hard hat

(C) respirator

(D) hearing protection

SOLUTIONS

1. Hand signals in the crane industry are standardized. This signal means to lower the boom and raise the load. If the fingers were not shown (a fist was being made), the signal would mean to lower the boom only [OSHA 29 CFR 1926.550(a)(4)].

The answer is (D).

2. OSHA does not normally consider an open trench to be a confined space. Although there are some combinations of base metals and rods that produce toxic fumes and would require respirators, these are rare in normal construction work, so welding in a trench would not normally result in the accumulation of toxic gases or the need for a respirator. A hard hat is always required [OSHA 29 CFR 1926.100].

The answer is (B).

Construction

58 Computer Software

1. Which of the following flowcharts does NOT represent a complete program?

(A)

(B)

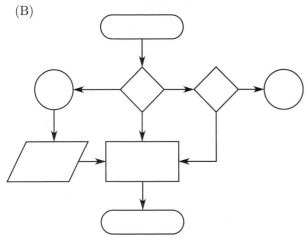

(C)

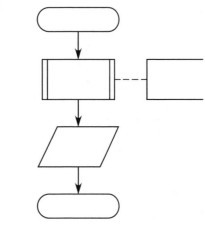

(D)

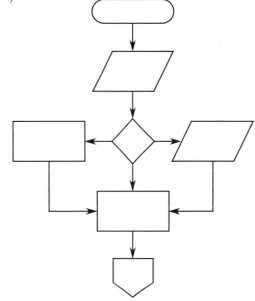

Computational Tools

2. What flowchart element would be used to represent an IF...THEN statement?

(A)

(B)

(C)

(D)

3. In programming, a recursive function is one that

(A) calls previously used functions

(B) generates functional code to replace symbolic code

(C) calls itself

(D) compiles itself in real time

4. Structured programming is to be used to determine whether examinees pass a test. A passing score is 70 or more out of a possible 100. Which of the following IF statements would set the variable PASSED to 1 (true) when the variable SCORE is passing, and set the variable PASSED to 0 (false) when the variable SCORE is not passing?

(A) IF SCORE > 70 PASSED = 1

    ELSE PASSED = 0

(B) IF SCORE > 69 PASSED = 1

    ELSE PASSED = 0

(C) IF SCORE < 69 PASSED = 1

(D) IF SCORE < 69 PASSED = 0

    ELSE PASSED = 1

5. A structured programming segment contains the following program segment. What is the value of Y after the segment is executed?

$$Y = 4$$
$$B = 4$$
$$Y = 3\,{}^*B - 6$$
$$\text{IF } Y > B \text{ THEN } Y = B - 2$$
$$\text{IF } Y < B \text{ THEN } Y = Y + 2$$
$$\text{IF } Y = B \text{ THEN } Y = B + 2$$

(A) 2

(B) 6

(C) 8

(D) 12

6. The structured programming segment shown implements which of the following equations? The variable N is an integer greater than 0.

$$A = X$$
$$\text{DO UNTIL } N = 0$$
$$Y = A\text{*}X$$
$$A = Y$$
$$N = N - 1$$
$$\text{END UNTIL}$$

(A) $Y = X!$

(B) $Y = X^{N+1}$

(C) $Y = X^{N-1}$

(D) $Y = X^N$

7. In the structured programming fragment shown, what can line 2 be accurately described as?

```
1   REAL X, Y
2   X = 3
3   Y = COS(X)
4   PRINT Y
```

(A) an assignment

(B) a command

(C) a declaration

(D) a function

Computational Tools

8. In a typical spreadsheet program, what cell is directly below cell AB4?

(A) AB5

(B) AC4

(C) AC5

(D) BC4

9. Which of the following terms is best defined as a formula or set of steps for solving a particular problem?

(A) program

(B) software

(C) firmware

(D) algorithm

10. Which of the following is the computer language that is executed within a computer's central processing unit?

(A) MS-DOS

(B) high-level language

(C) assembly language

(D) machine language

11. Which of the following best defines a compiler?

(A) hardware that is used to translate high-level language to machine code

(B) software that collects and stores executable commands in a program

(C) software that is used to translate high-level language into machine code

(D) hardware that collects and stores executable commands in a program

12. The effect of using recursive functions in a program is generally to

(A) use less code and less memory

(B) use less code and more memory

(C) use more code and less memory

(D) use more code and more memory

13. When can an 8-bit system correctly access more than 128 different integers?

(A) when the integers are in the range of $[-255, 0]$

(B) when the integers are in the range of $[0, 256]$

(C) when the integers are in the range of $[-128, 128]$

(D) when the integers are in the range of $[0, 512]$

14. What computer operating system (OS) is required to view a document saved in HTML (hypertext markup language) format?

I.   Apple OS

II. MS-DOS

III. Windows

IV. Unix

(A) I or III

(B) I, II, or III

(C) I, III, or IV

(D) I, II, III, or IV

15. A typical spreadsheet for economic evaluation of alternatives uses cell F4 to store the percentage value of inflation rate. The percentage rate is assumed to be constant throughout the lifetime of the study. What variable should be used to access that value throughout the model?

(A) F4

(B) $F4

(C) %F4

(D) F4

16. Refer to the following portion of a spreadsheet.

	A	B	C
1	10	11	12
2	1	A2^2	
3	2	A3^2	
4	3	A4^2	
5	4	A5^2	

The top-to-bottom values in column B will be

(A) 11, 1, 2, 3, 4

(B) 11, 1, 3, 6, 10

(C) 11, 1, 4, 9, 16

(D) 11, 1, 5, 12, 22

17. Refer to the following portion of a spreadsheet.

	A	B	C	D
1	10	11	12	13
2	5	4	B2*A$1	
3	6	5	B3*B$1	
4	7	6	B4*C$1	
5	8	7	B5*D$1	

The top-to-bottom values in column C will be

(A) 12, 20, 30, 42, 56

(B) 12, 40, 55, 72, 91

(C) 12, 50, 66, 84, 104

(D) 12, 100, 121, 144, 169

SOLUTIONS

1. A flowchart must begin and end with a terminal symbol. The symbol at the bottom of answer D is the "off-page" symbol, which indicates that the flowchart continues on the next page. This is not a complete program.

The answer is (D).

2. An IF...THEN statement alters the flow of a program based on some criterion that can be evaluated as true or false. The diamond-shaped symbol is used to represent a decision.

The answer is (B).

3. A recursive function calls itself.

The answer is (C).

4. Answer A will not give the correct response when SCORE = 70. Answer C will not set PASSED to 0. Answer D will not give the correct response when SCORE = 69. Answer B will set PASSED to 1 for SCORE = 70 to 100 and will set PASSED to 0 for SCORE = 0 to 69.

The answer is (B).

5. The first operation changes the value of Y.

$$Y = 3 \times 4 - 6 = 6$$

The first IF statement is satisfied, so the operation is performed.

$$Y = 4 - 2 = 2$$

However, the program execution does not end here.

The value of Y is then less than B, so the second IF statement is executed. This statement is satisfied, so the operation is performed.

$$Y = 2 + 2 = 4$$

The value of Y is then equal to B, so the third IF statement is executed. This statement is satisfied, so the operation is performed.

$$Y = 4 + 2 = 6$$

The answer is (B).

6. The DO loop will be executed N times. After the first execution, $Y = X^2$. After each subsequent execution of the loop, Y is multiplied by X. Therefore, the segment calculates $Y = X^{N+1}$.

The answer is (B).

7. Line 2 is an assignment. A command directs the computer to take some action, such as PRINT. A declaration states what type of data a variable will contain (like REAL) and reserves space for it in memory. A function performs some specific operation (like finding the COSine of a number) and returns a value to the program.

The answer is (A).

8. Spreadsheets generally label a cell by giving its column and row, in that order. Cell AB4 is in column AB, row 4. The cell directly below AB4 is in column AB, row 5, designated as AB5.

The answer is (A).

9. A program is a sequence of instructions that implements a formula or set of steps but is not a formula or set of steps. Software and firmware are programs on media. An algorithm is a formula or set of steps for solving a particular problem that is often implemented as a program.

The answer is (D).

10. The central processing unit executes a version of the program that has been compiled into the machine language and that is in the form of operations and operands specific to the machine's coding.

The answer is (D).

11. A compiler is a program (i.e., software) that converts programs written in higher-level languages to lower-level languages that the computer can understand.

The answer is (C).

12. A recursive function calls itself. Since the function does not need to be coded in multiple places, less code is used. Each subsequent call of the function must be carried out in a different location, so more memory is used.

The answer is (B).

13. An 8-bit system can represent $(2)^8 = 256$ different distinct integers. Normally, the 8th bit is used for the sign, and only 7 bits are used for magnitude, resulting in a range of $[-127, 128]$ or $[-128, 127]$ (counting zero as one of the integers). If all of the integers are known or assumed to have the same sign, a range of 256 integers is available. All of the answer choices except A contain more than 256 distinct integers.

The answer is (A).

14. HTML may be viewed on any computer with a compatible browser.

The answer is (D).

15. The dollar sign symbol, "$", is used in spreadsheets to "fix" the column and/or row designator following it when other columns or rows are permitted to vary.

The answer is (D).

16. Except for the first entry (which is 11), column B calculates the square of the values in column A. The entries are 11, $(1)^2$, $(2)^2$, $(3)^2$, and $(4)^2$.

The answer is (C).

17. Except for the first entry (which is 12), column C is found by taking the numbers from column B and then multiplying by the entries in row 1. For example, B2 * A$1 means to multiply the entry in B2, given as 4 in the problem statement, by the number entered in cell A1, which is 10. This product is 40.

The entries are 12, 4×10, 5×11, 6×12, 7×13, or 12, 40, 55, 72, 91.

The answer is (B).

59 Engineering Economics

PRACTICE PROBLEMS

1. Permanent mineral rights on a parcel of land are purchased for an initial lump-sum payment of $100,000. Profits from mining activities are $12,000 each year, and these profits are expected to continue indefinitely. The interest rate earned on the initial investment is most nearly

(A) 8.3%

(B) 9.0%

(C) 10%

(D) 12%

2. $1000 is deposited in a savings account that pays 6% annual interest, and no money is withdrawn for three years. The account balance after three years is most nearly

(A) $1120

(B) $1190

(C) $1210

(D) $1280

3. An oil company is planning to install a new 80 mm pipeline to connect storage tanks to a processing plant 1500 m away. The connection will be needed for the foreseeable future. An annual interest rate of 8% is assumed, and annual maintenance and pumping costs are considered to be paid in their entireties at the end of the years in which their costs are incurred.

initial cost	$1500
service life	12 yr
salvage value	$200
annual maintenance	$400
pump cost/hour	$2.50
pump operation	600 hr/yr

The capitalized cost of running and maintaining the 80 mm pipeline is most nearly

(A) $15,000

(B) $20,000

(C) $24,000

(D) $27,000

4. New 200 mm diameter pipeline is installed over a distance of 1000 m. Annual maintenance and pumping costs are considered to be paid in their entireties at the end of the years in which their costs are incurred. The pipe has the following costs and properties.

initial cost	$1350
annual interest rate	6%
service life	6 yr
salvage value	$120
annual maintenance	$500
pump cost/hour	$2.75
pump operation	2000 hr/yr

What is most nearly the equivalent uniform annual cost (EUAC) of the pipe?

(A) $5700

(B) $5900

(C) $6100

(D) $6300

5. New 120 mm diameter pipeline is installed over a distance of 5000 m. Annual maintenance and pumping costs are considered to be paid in their entireties at the end of the years in which their costs are incurred. The pipe has the following costs and properties.

initial cost	$2500
annual interest rate	10%
service life	12 yr
salvage value	$300
annual maintenance	$300
pump cost/hour	$1.40
pump operation	600 hr/yr

What is most nearly the equivalent uniform annual cost (EUAC) of the pipe?

(A) $1200

(B) $1300

(C) $1400

(D) $1500

6. A construction company purchases 100 m of 40 mm diameter steel cable with an initial cost of $4500. The annual interest rate is 4%, and annual maintenance costs are considered to be paid in their entireties at the end of the years in which their costs are incurred. The annual maintenance cost of the cable is $200/yr over a service life of nine years. Using Modified Accelerated Cost Recovery System (MACRS) depreciation and assuming a seven year recovery period, the depreciation allowance for the cable in the first year of operation is most nearly

(A) $640

(B) $670

(C) $720

(D) $860

7. A piece of equipment has an initial cost of $5000 in year 1. The maintenance cost is $300/yr for the total lifetime of seven years. During years 1–3, the rate of inflation is 5%, and the effective annual rate of interest is 9%. The uninflated present worth of the equipment during year 1 is most nearly

(A) $3200

(B) $3300

(C) $3400

(D) $3500

8. A company is considering buying a computer with the following costs and interest rate.

initial cost	$3900
salvage value	$1800
useful life	10 years
annual maintenance	$390
interest rate	6%

The equivalent uniform annual cost (EUAC) of the computer is most nearly

(A) $740

(B) $780

(C) $820

(D) $850

9. A computer with a useful life of 13 years has the following costs and interest rate.

initial cost	$5500
salvage value	$3100
annual maintenance	
years 1–8	$275
years 9–13	$425
interest rate	6%

The equivalent uniform annual cost (EUAC) of the computer is most nearly

(A) $730

(B) $780

(C) $820

(D) $870

10. A computer with an initial cost of $1500 and an annual maintenance cost of $500/yr is purchased and kept indefinitely without any change in its annual maintenance costs. The interest rate is 4%. The present worth of all expenditures is most nearly

(A) $12,000

(B) $13,000

(C) $14,000

(D) $15,000

11. A computer with a useful life of 12 years has an initial cost of $2300 and a salvage value of $350. The interest rate is 6%. Using the straight line method, the total depreciation of the computer for the first five years is most nearly

(A) $760

(B) $810

(C) $830

(D) $920

12. A computer with a useful life of 12 years has an initial cost of $3200 and a salvage value of $100. The interest rate is 10%. Using the Modified Accelerated Cost Recovery System (MACRS) method of depreciation and a 10 year recovery period, what is most nearly the book value of the computer after the second year?

(A) $1900

(B) $2100

(C) $2300

(D) $2400

13. A computer with a useful life of five years has an initial cost of $6000. The salvage value is $2300, and the annual maintenance is $210/yr. The interest rate is 8%. What is most nearly the present worth of the costs for the computer?

(A) $5200

(B) $5300

(C) $5600

(D) $5700

14. A company must purchase a machine that will be used over the next eight years. The purchase price is $10,000, and the salvage value after eight years is $1000. The annual insurance cost is 2% of the purchase price, the electricity cost is $300 per year, and maintenance and replacement parts cost $100 per year. The effective annual interest rate is 6%. Neglect taxes. The effective uniform annual cost (EUAC) of ownership is most nearly

(A) $1200

(B) $2100

(C) $2200

(D) $2300

15. A company purchases a piece of equipment for $15,000. After nine years, the salvage value is $900. The annual insurance cost is 5% of the purchase price, the electricity cost is $600/yr, and the maintenance and replacement parts cost is $120/yr. The effective annual interest rate is 10%. Neglecting taxes, what is most nearly the present worth of the equipment if it is expected to save the company $4500 per year?

(A) $2300

(B) $2800

(C) $3200

(D) $3500

SOLUTIONS

1. Use the capitalized cost equation to find the interest rate earned.

$$P = \frac{A}{i}$$

$$i = \frac{A_{\text{profit}}}{P_{\text{cost}}} = \frac{\$12,000}{\$100,000}$$

$$= 0.12 \quad (12\%)$$

The answer is (D).

2. Find the future worth of $1000.

$$F = P(1+i)^n = (\$1000)(1 + 0.06)^3$$

$$= \$1191 \quad (\$1190)$$

The answer is (B).

3. The annual cost of running and maintaining the 80 mm pipeline is

$$A = \left(\frac{\$2.50}{\text{hr}}\right)(600 \text{ hr}) + \$400 = \$1900$$

Capitalized costs are the present worth of an infinite cash flow.

$$P = \frac{A}{i} = \frac{\$1900}{0.08}$$

$$= \$23,750 \quad (\$24,000)$$

The answer is (C).

4. The equivalent uniform annual cost (EUAC) is the uniform annual amount equivalent of all cash flows. When calculating the EUAC, costs are positive and income is negative.

$$
\begin{aligned}
\text{EUAC}_{80} &= A_{\text{initial}} + A_{\text{maintenance}} \\
&\quad + A_{\text{pump}} - A_{\text{salvage}} \\
&= (\$1350)(A/P, 6\%, 6) + \$500 \\
&\quad + \left(\frac{\$2.75}{\text{hr}}\right)(2000 \text{ hr}) \\
&\quad - (\$120)(A/F, 6\%, 6) \\
&= (\$1350)(0.2034) + \$500 \\
&\quad + \$5500 - (\$120)(0.1434) \\
&= \$6257 \quad (\$6300)
\end{aligned}
$$

The answer is (D).

Engineering Economics

5. The equivalent uniform annual cost (EUAC) is the uniform annual amount equivalent of all cash flows. When calculating the EUAC, costs are positive and income is negative.

$$\begin{aligned}
\text{EUAC}_{120} &= A_{\text{initial}} + A_{\text{maintenance}} \\
&\quad + A_{\text{pump}} - A_{\text{salvage}} \\
&= (\$2500)(A/P, 10\%, 12) + \$300 \\
&\quad + \left(\frac{\$1.40}{\text{hr}}\right)(600 \text{ hr}) \\
&\quad - (\$300)(A/F, 10\%, 12) \\
&= (\$2500)(0.1468) + \$300 + \$840 \\
&\quad - (\$300)(0.0468) \\
&= \$1492.96 \quad (\$1500)
\end{aligned}$$

The answer is (D).

6. The MACRS factor for the first year, given a 7 year recovery period, is 14.29%.

$$\begin{aligned}
D_1 &= (\text{factor})\,C \\
&= (0.1429)(\$4500) \\
&= \$643 \quad (\$640)
\end{aligned}$$

The answer is (A).

7. If the unadjusted interest rate is used to calculate the present worth, the answer will be in dollars affected by three years of inflation. To find the uninflated worth three years ago, the effect of inflation during those years must be eliminated from the calculation. To find the answer in uninflated dollars, determine the interest rate adjusted for inflation.

$$\begin{aligned}
d &= i + f + (i \times f) \\
&= 0.09 + 0.05 + (0.09)(0.05) \\
&= 0.1445
\end{aligned}$$

Use this adjusted rate in the single payment present worth equation, substituting d for i.

$$\begin{aligned}
P &= F(1+d)^{-n} \\
&= (\$5000)(1 + 0.1445)^{-3} \\
&= \$3335 \quad (\$3300)
\end{aligned}$$

The answer is (B).

8. The equivalent uniform annual cost (EUAC) is the uniform annual amount equivalent to all cash flows. When calculating the EUAC, costs are positive and income is negative.

$$\begin{aligned}
\text{EUAC} &= A_{\text{initial}} + A_{\text{maintenance}} - A_{\text{salvage}} \\
&= (\$3900)(A/P, 6\%, 10) + \$390 \\
&\quad - (\$1800)(A/F, 6\%, 10) \\
&= (\$3900)(0.1359) + \$390 \\
&\quad - (\$1800)(0.0759) \\
&= \$783 \quad (\$780)
\end{aligned}$$

The answer is (B).

9. The equivalent uniform annual cost (EUAC) is the uniform annual amount equivalent to all cash flows. When calculating the EUAC, costs are positive and income is negative. An expedient way to find the annual worth of the maintenance for the computer is to divide the maintenance costs into two annual series, one of $275 lasting from year 1 to year 13, and one of $150 (the difference between $425 and $275) lasting from year 9 to year 13. Find the future value in year 13 for each series, add them, and then convert the result back into a single annual amount.

$$\begin{aligned}
F_{\$275} &= A(F/A, 6\%, 13) = (\$275)(18.8821) \\
&= \$5192.60 \\
F_{\$150} &= A(F/A, 6\%, 5) = (\$150)(5.6371) \\
&= \$845.60 \\
F_{\text{maintenance}} &= F_{\$275} + F_{\$150} = \$5192.60 + \$845.60 \\
&= \$6038 \\
A_{\text{maintenance}} &= F_{\text{maintenance}}(A/F, 6\%, 13) \\
&= (\$6038)(0.0530) \\
&= \$320
\end{aligned}$$

Now calculate the EUAC.

$$\begin{aligned}
\text{EUAC} &= A_{\text{initial}} + A_{\text{maintenance}} - A_{\text{salvage}} \\
&= (\$5500)(A/P, 6\%, 13) + \$320 \\
&\quad - (\$3100)(A/F, 6\%, 13) \\
&= (\$5500)(0.1130) + \$320 \\
&\quad - (\$3100)(0.0530) \\
&= \$777 \quad (\$780)
\end{aligned}$$

The answer is (B).

10. The expenditures for the computer are the initial cost of $1500 and the annual maintenance cost of $500. The annual costs continue indefinitely, so find the present worth of an infinite cash flow.

$$P_\text{maintenance} = \frac{A}{i} = \frac{\$500}{0.04} = \$12,500$$

The present worth of all expenditures is

$$P_\text{total} = P_\text{initial} + P_\text{annual} = \$1500 + \$12,500$$
$$= \$14,000$$

The answer is (C).

11. With the straight line method, the depreciation is the same every year. Find the annual depreciation.

$$D_j = \frac{C - S_n}{n} = \frac{\$2300 - \$350}{12} = \$162.50$$

The total depreciation for five years is

$$\sum D_{1-5} = (5)(\$162.50) = \$812.50 \quad (\$810)$$

The answer is (B).

12. Subtract the first two years' depreciation from the original cost.

year	factor (%)	D_j
1	10.00	$(0.10)(\$3200) = \$320$
2	18.00	$(0.18)(\$3200) = \underline{\$576}$
		$\sum D_j = \$896$

The book value is

$$\text{BV} = \text{initial cost} - \sum D_j = \$3200 - \$896$$
$$= \$2304 \quad (\$2300)$$

The answer is (C).

13. Bring all costs and benefits into the present.

$$P_\text{total} = P_\text{initial} + P_\text{maintenance} - P_\text{salvage}$$
$$= \$6000 + (\$210)(P/A, 8\%, 5)$$
$$\quad - (\$2300)(P/F, 8\%, 5)$$
$$= \$6000 + (\$210)(3.9927)$$
$$\quad - (\$2300)(0.6806)$$
$$= \$5273 \quad (\$5300)$$

The answer is (B).

14. The effective uniform annual cost (EUAC) is the annual cost equivalent of all costs. When calculating the EUAC, costs are positive and income is negative. Find the annual equivalents of all costs and add them together to get the EUAC.

$$\text{EUAC} = C_\text{initial}(A/P, 6\%, 8)$$
$$\quad + A_\text{electricity} + A_\text{maintenance}$$
$$\quad + A_\text{insurance} - S_8(A/F, 6\%, 8)$$
$$= (\$10,000)(0.1610) + \$300 + \$100$$
$$\quad + (0.02)(\$10,000)$$
$$\quad - (\$1000)(0.1010)$$
$$= \$2109 \quad (\$2100)$$

The answer is (B).

15. Add the present worths of all cash flows.

$$P_\text{total} = -C_\text{initial} - A_\text{electricity}(P/A, 10\%, 9)$$
$$\quad - A_\text{maintenance}(P/A, 10\%, 9)$$
$$\quad - A_\text{insurance}(P/A, 10\%, 9)$$
$$\quad + A_\text{benefits}(P/A, 10\%, 9)$$
$$\quad + S_9(P/F, 10\%, 9)$$
$$= -\$15,000 - (\$600)(5.7590)$$
$$\quad - (\$120)(5.7590)$$
$$\quad - (0.05)(\$15,000)(5.7590) + (\$4500)(5.7590)$$
$$\quad + (\$900)(0.4241)$$
$$= \$2831 \quad (\$2800)$$

The answer is (B).

60 Professional Practice

PRACTICE PROBLEMS

1. What must be proven for damages to be collected from a strict liability in tort?

(A) that willful negligence caused an injury

(B) that willful or unwillful negligence caused an injury

(C) that the manufacturer knew about a product defect before the product was released

(D) none of the above

2. A material breach of contract occurs when the

(A) contractor uses material not approved by the contract for use

(B) contractor's material order arrives late

(C) owner becomes insolvent

(D) contractor installs a feature incorrectly

3. If a contract has a value engineering clause and a contractor suggests to the owner that a feature or method be used to reduce the annual maintenance cost of the finished project, what will be the most likely outcome?

(A) The contractor will be able to share one time in the owner's expected cost savings.

(B) The contractor will be paid a fixed amount (specified by the contract) for making a suggestion, but only if the suggestion is accepted.

(C) The contract amount will be increased by some amount specified in the contract.

(D) The contractor will receive an annuity payment over some time period specified in the contract.

4. A tort is

(A) a civil wrong committed against another person

(B) a section of a legal contract

(C) a legal procedure in which complaints are heard in front of an arbitrator rather than a judge or jury

(D) the breach of a contract

5. If a contract does not include the boilerplate clause, "Time is of the essence," which of the following is true?

(A) It is difficult to recover losses for extra hours billed.

(B) Standard industry time guidelines apply.

(C) Damages for delay cannot be claimed.

(D) Workers need not be paid for downtime in the project.

6. Which statement is true regarding the legality and enforceability of contracts?

(A) For a contract to be enforceable, it must be in writing.

(B) A contract to perform illegal activity will still be enforced by a court.

(C) A contract must include a purchase order.

(D) Mutual agreement of all parties must be evident.

7. Which option best describes the contractual lines of privity between parties in a general construction contract?

(A) The consulting engineer will have a contractual obligation to the owner, but will not have a contractual obligation with the general contractor or the subcontractors.

(B) The consulting engineer will have a contractual obligation to the owner and the general contractor.

(C) The consulting engineer will have a contractual obligation to the owner, general contractor, and subcontractors.

(D) The consulting engineer will have a contractual obligation to the general contractor, but will not have a contractual obligation to the owner or subcontractors.

8. A contract has a value engineering clause that allows the parties to share in improvements that reduce cost. The contractor had originally planned to transport concrete on-site for a small pour with motorized wheelbarrows. On the day of the pour, however, a concrete pump is available and is used, substantially reducing the contractor's labor cost for the day. This is an example of

(A) value engineering whose benefit will be shared by both contractor and owner

(B) efficient methodology whose benefit is to the contractor only

(C) value engineering whose benefit is to the owner only

(D) cost reduction whose benefit will be shared by both contractor and laborers

9. In which of the following fee structures is a specific sum paid to the engineer for each day spent on the project?

(A) salary plus

(B) per-diem fee

(C) lump-sum fee

(D) cost plus fixed fee

10. What type of damages is paid when responsibility is proven but the injury is slight or insignificant?

(A) nominal

(B) liquidated

(C) compensatory

(D) exemplary

SOLUTIONS

1. In order to prove strict liability in tort, it must be shown that a product defect caused an injury. Negligence need not be proven, nor must the manufacturer know about the defect before release.

The answer is (D).

2. A material breach of the contract is a significant event that is grounds for cancelling the contract entirely. Typical triggering events include failure of the owner to make payments, the owner causing delays, the owner declaring bankruptcy, the contractor abandoning the job, or the contractor getting substantially off schedule.

The answer is (C).

3. Changes to a structure's performance, safety, appearance, or maintenance that benefit the owner in the long run will be covered by the value engineering clause of a contract. Normally, the contractor is able to share in cost savings in some manner by receiving a payment or credit to the contract.

The answer is (A).

4. A tort is a civil wrong committed against a person or his/her property which results in some form of damages. Torts are normally resolved through lawsuits.

The answer is (A).

5. This clause must be included in order to recover damages due to delay.

The answer is (C).

6. In order for a contract to be legally binding, it must

- be established for a legal purpose

- contain a mutual agreement by all parties

- have consideration, or an exchange of something of value (e.g., a service is provided in exchange for a fee)

- not obligate parties to perform illegal activity

- not be between parties that are mentally incompetent, minors, or do not otherwise have the power to enter into the contract

A contract does not need to use as its basis or include a purchase order to be enforceable. Oral contracts may be legally binding in some instances, depending on the circumstances and purpose of the contract. Oral contracts may be difficult to enforce, however, and should not be used for engineering and construction agreements.

The answer is (D).

7. With a general construction contract, a consulting engineer will be hired by the owner to develop the design and contract documents, as well as to assist in the preparation of the bid documents and provide contract administrative services during the construction phase. The contract documents produced by the engineer will form the basis of the owner's agreement with the contractor. Although the engineer will work closely with the contractor during the construction phase, and may work with subcontractors as well, the engineer will not have a contractual line of privity with either party.

The answer is (A).

8. The problem gives an example of efficient methodology, where the benefit is to the contractor only. It is not an example of value engineering, as the change affects the contractor, not the owner. Performance, safety, appearance, and maintenance are unaffected.

The answer is (B).

9. A specific fee is paid to the engineer for each day on the job in a per-diem fee structure.

The answer is (B).

10. Nominal damages are awarded for inconsequential injuries.

The answer is (A).

Ethics/
Prof. Prac.

61 Ethics

PRACTICE PROBLEMS

1. An environmental engineer with five years of experience reads a story in the daily paper about a proposal being presented to the city council to construct a new sewage treatment plant near protected wetlands. Based on professional experience and the facts presented in the newspaper, the engineer suspects the plant would be extremely harmful to the local ecosystem. Which of the following would be an acceptable course of action?

(A) The engineer should contact appropriate agencies to get more data on the project before making a judgment.

(B) The engineer should write an article for the paper's editorial page urging the council not to pass the project.

(C) The engineer should circulate a petition through the community condemning the project, and present the petition to the council.

(D) The engineer should do nothing because he doesn't have enough experience in the industry to express a public opinion on the matter.

2. An engineer is consulting for a construction company that has been receiving bad publicity in the local papers about its waste-handling practices. Knowing that this criticism is based on public misperceptions and the paper's thirst for controversial stories, the engineer would like to write an article to be printed in the paper's editorial page. What statement best describes the engineer's ethical obligations?

(A) The engineer's relationship with the company makes it unethical for him to take any public action on its behalf.

(B) The engineer should request that a local representative of the engineering registration board review the data and write the article in order that an impartial point of view be presented.

(C) As long as the article is objective and truthful, and presents all relevant information including the engineer's professional credentials, ethical obligations have been satisfied.

(D) The article must be objective and truthful, pres“ent all relevant information including the engineer's professional credentials, and disclose all details of the engineer's affiliation with the company.

3. After making a presentation for an international project, an engineer is told by a foreign official that his company will be awarded the contract, but only if it hires the official's brother as an advisor to the project. The engineer sees this as a form of extortion and informs his boss. His boss tells him that, while it might be illegal in the United States, it is a customary and legal business practice in the foreign country. The boss impresses upon the engineer the importance of getting the project, but leaves the details up to the engineer. What should the engineer do?

(A) He should hire the official's brother, but insist that he perform some useful function for his salary.

(B) He should check with other companies doing business in the country in question, and if they routinely hire relatives of government officials to secure work, then he should do so too.

(C) He should withdraw his company from consideration for the project.

(D) He should inform the government official that his company will not hire the official's brother as a precondition for being awarded the contract, but invite the brother to submit an application for employment with the company.

4. If one is aware that a registered engineer willfully violates a state's rule of professional conduct, one should

(A) do nothing

(B) report the violation to the state's engineering registration board

(C) report the violation to the employer

(D) report the violation to the parties it affects

5. Which of the following is an ethics violation specifically included in the NCEES *Model Rules*?

(A) an engineering professor "moonlighting" as a private contractor

(B) an engineer investing money in the stock of the company for which he/she works

(C) a civil engineer with little electrical experience signing the plans for an electric generator

(D) none of the above

6. A senior licensed professional engineer with 30 years of experience in geotechnical engineering is placed in charge of a multidisciplinary design team consisting of a structural group, a geotechnical group, and an environmental group. In this role, the engineer is responsible for supervising and coordinating the efforts of the groups when working on large interconnected projects. In order to facilitate coordination, designs are prepared by the groups under the direct supervision of the group leader, and then they are submitted to her for review and approval. This arrangement is ethical as long as

(A) the engineer signs and seals each design segment only after being fully briefed by the appropriate group leader

(B) the engineer signs and seals only those design segments pertaining to geotechnical engineering

(C) each design segment is signed and sealed by the licensed group leader responsible for its preparation

(D) the engineer signs and seals each design segment only after it has been reviewed by an independent consulting engineer who specializes in the field in which it pertains

7. The National Society of Professional Engineers' (NSPE) Code of Ethics addresses competitive bidding. Which of the following is NOT stipulated?

(A) Engineers and their firms may refuse to bid competitively on engineering services.

(B) Clients are required to seek competitive bids for design services.

(C) Federal laws governing procedures for procuring engineering services (e.g., competitive bidding) remain in full force.

(D) Engineers and their societies may actively lobby for legislation that would prohibit competitive bidding for design services.

8. You are a city engineer in charge of receiving bids on behalf of the city council. A contractor's bid arrives with two tickets to a professional football game. The bid is the lowest received. What should you do?

(A) Return the tickets and accept the bid.

(B) Return the tickets and reject the bid.

(C) Discard the tickets and accept the bid.

(D) Discard the tickets and reject the bid.

9. A relatively new engineering firm is considering running an advertisement for their services in the local newspaper. An ad agency has supplied them with four concepts. Of the four ad concepts, which one(s) would be acceptable from the standpoint of professional ethics?

I.   an advertisement contrasting their successes over the past year with their nearest competitors' failures

II.   an advertisement offering a free television to anyone who hires them for work valued at over $10,000

III. an advertisement offering to beat the price of any other engineering firm for the same services

IV.   an advertisement that tastefully depicts their logo against the backdrop of the Golden Gate Bridge

(A) I and III

(B) I, III, and IV

(C) II, III, and IV

(D) neither I, II, III, nor IV

10. Complete the sentence: "A professional engineer who took the licensing examination in mechanical engineering

(A) may not design in electrical engineering."

(B) may design in electrical engineering if she feels competent."

(C) may design in electrical engineering if she feels competent and the electrical portion of the design is insignificant and incidental to the overall job."

(D) may design in electrical engineering if another engineer checks the electrical engineering work."

11. An engineering firm is hired by a developer to prepare plans for a shopping mall. Prior to the final bid date, several contractors who have received bid documents and plans contact the engineering firm with requests for information relating to the project. What can the engineering firm do?

(A) The firm can supply requested information to the contractors as long as it does so fairly and evenly. It cannot favor or discriminate against any contractor.

(B) The firm should supply information to only those contractors that it feels could safely and economically perform the construction services.

(C) The firm cannot reveal facts, data, or information relating to the project that might prejudice a contractor against submitting a bid on the project.

(D) The firm cannot reveal facts, data, or information relating to the project without the consent of the client as authorized or required by law.

SOLUTIONS

1. The engineer certainly has more experience and knowledge in the field than the general public or even the council members who will have to vote on the issue. Therefore, the engineer is qualified to express his opinion if he wishes to do so. Before the engineer takes any public position, however, the engineer is obligated to make sure that all the available information has been collected.

The answer is (A).

2. It is ethical for the engineer to issue a public statement concerning a company he works for, provided he makes that relationship clear and provided the statement is truthful and objective.

The answer is (D).

3. Hiring the official's brother as a precondition for being awarded the contract is a form of extortion. Depending on the circumstances, however, it may be legal to do so according to U.S. law. (The Foreign Corrupt Practices Act of 1977 allows American companies to pay extortion in some cases.) This practice, however, is not approved by the NCEES *Model Rules*:

> Registrants shall not offer, give, solicit, or receive, either directly or indirectly, any commission or gift, or other valuable consideration in order to secure work.

The answer is (D).

4. A violation should be reported to the organization that has promulgated the rule.

The answer is (B).

5. The NCEES *Model Rules* specifically states that registrants may not perform work beyond their level of expertise. The other two examples may be unethical under some circumstances, but are not specifically forbidden by the NCEES code.

The answer is (C).

6. According to the NCEES *Model Rules*,

> Licensees may accept assignments for coordination of an entire project, provided that each design segment is signed and sealed by the registrant responsible for preparation of that design segment.

The answer is (C).

7. Clients are not required to seek competitive bids. In fact, many engineering societies discourage the use of bidding to procure design services because it is believed that competitive bidding results in lower-quality construction.

The answer is (B).

8. Registrants should not accept gifts from parties expecting special consideration, so the tickets cannot be kept. They also should not be merely discarded, for several reasons. Inasmuch as the motive of the contractor is not known with certainty, in the absence of other bidding rules, the bid may be accepted.

The answer is (A).

9. None of the ads is acceptable from the standpoint of professional ethics. Concepts I and II are explicitly prohibited by the NCEES *Model Rules*. Concept III demeans the profession of engineering by placing the emphasis on price as opposed to the quality of services. Concept IV is a misrepresentation; the picture of the Golden Gate Bridge in the background might lead some potential clients to believe that the engineering firm in question had some role in the design or construction of that project.

The answer is (D).

10. Although the laws vary from state to state, engineers are usually licensed generically. Engineers are licensed as "professional engineers." The scope of their work is limited only by their competence. In the states where the license is in a particular engineering discipline, an engineer may "touch upon" another discipline when the work is insignificant and/or incidental.

The answer is (C).

11. It is normal for engineers and architects to clarify the bid documents. However, some information may be proprietary to the developer. The engineering firm should only reveal information that has already been publicly disseminated or approved for release with the consent of the client.

The answer is (D).

62 Licensure

PRACTICE PROBLEMS

There are no problems in this book covering the subject of licensure.